JAVIER ROBLES

Incidence of hypochlorhydria and nitrites in the gastric mucosa.

JAVIER ROBLES

Incidence of hypochlorhydria and nitrites in the gastric mucosa.

As a trigger for atrophy and metaplasia.

Imprint

Any brand names and product names mentioned in this book are subject to trademark, brand or patent protection and are trademarks or registered trademarks of their respective holders. The use of brand names, product names, common names, trade names, product descriptions etc. even without a particular marking in this work is in no way to be construed to mean that such names may be regarded as unrestricted in respect of trademark and brand protection legislation and could thus be used by anyone.

Cover image: www.ingimage.com

This book is a translation from the original published under ISBN 978-613-9-40482-7.

Publisher:
Sciencia Scripts
is a trademark of
Dodo Books Indian Ocean Ltd. and OmniScriptum S.R.L publishing group

120 High Road, East Finchley, London, N2 9ED, United Kingdom
Str. Armeneasca 28/1, office 1, Chisinau MD-2012, Republic of Moldova, Europe
Printed at: see last page
ISBN: 978-620-8-08443-1

DEDICATION

GOD THE HEAVENLY FATHER for all your blessings. To María Angélica Calderón my beloved mother.

To Isaías and Narcisa, your memories will live on in my mind and heart.

To Laura, Sara, Marcial, Martha and Teresa my dear siblings.

THANK YOU

I would like to express my most sincere gratitude to Dr. Miguel Bilbao Díaz, Thesis Tutor, to my long-time friend José Guevara, to Dr. Fabián Romero, who is the subject manager and gastroenterologist at the Hospital, without his help it would not have been possible to complete this work, to Mrs. Rosemery Velasteguí for her unconditional collaboration, to Paulo Terán for his encouragement, to Lorena Peralta for her time spent and to all the people who supported me in the completion of this work.

SUMMARY

Modern medicine, both worldwide and nationally, has experienced in recent years important advances in the diagnosis of specific pathologies. One of the areas that has benefited most from these technological advances is gastroenterology and specifically upper and lower gastrointestinal endoscopy. The objective of this work was to establish the incidence of hypochlorhydria or achlorhydria and nitrites in the gastric mucosa in patients attended in the Gastroenterology Area of the Solca Riobamba Hospital as a trigger for Gastric Atrophy and Metaplasia, using as a method the upper digestive endoscopy, the analysis of gastric juice and later the biopsy obtained in this procedure. The sample consisted of 105 patients who underwent endoscopic examinations, where the importance of this examination and the usefulness of taking a biopsy in this type of lesions was determined. The relationship between the macroscopic findings and the histopathological result was also determined, as well as the most frequent type of inflammatory infiltrate diagnosed, as well as the pathologies found, all of the above characterized according to sex, age, origin and educational level. It was determined that of the total number of patients seen, 36 had presence of nitrites in the gastric juice for 34.28 %. Hypochlorhydria was observed in 42 patients for 40 % of the total. It was found that in 36 patients there was a relationship of achlorhydria and nitrites together with metaplasia and atrophy, being 44.44 % of the female sex and 55.55 % of the male sex. It was also determined that all of them presented metaplasia, decrease of glandular groups and both pathologies, being these considered as important lesions for the development of a gastric carcinoma. 100% of the patients who underwent this procedure presented clinical signs of gastrointestinal disease, ranging from mild gastropathy to advanced stage neoplastic infiltrates. It can be concluded that both the presence of nitrites and achlorhydria are fundamental risk factors for the triggering of gastric atrophy and metaplasia.

KEYWORDS: ACHLORHYDRIA, -ATROPHY -ENDOSCOPY - UPPER GASTROINTESTINAL ENDOSCOPY -METAPLASIA,

ABSTRACT

The modern medicine at both global and national levels has experienced in recent years important advances in the diagnosis of specific pathologies. One areas that has benefited most from these technological advances is the gastroenterology and specifically the lower and upper endoscopy. The objective of this work was to establish the incidence of hypoachlorhydria or achlorhydria and nitrites in the gastric mucosa in patients in Gastroenterology area of Solca Hospital in Riobamba city as the caused of gastric atrophy and metaplasia, using the method the upper gastrointestinal endoscopy, the analysis of gastric juice and subsequently the biopsy samples obtained in this procedure. The sample was 105 patients who underwent endoscopic examinations, where it was determined the importance of this test and the utility of biopsy in this type of injury. We also determined the relationship between the macroscopic findings and the histopathological result, as well as the type of inflammatory infiltrate more frequently diagnosed, as also the pathologies encountered, all the above characterized by sex, age, origin and educational level. It was determined that the total number of patients treated, 36 were presented of nitrites in the gastric juice for a 34.28 %. Hypochlorhydria was observed in 42 patients for a 40 per cent of the total. It was found that in 36 patients there was a relationship of achlorhydria and nitrites jointly with atrophy and metaplasia, being the 44.44 % of females and the 55.55 % of male sex. Also it was determined that all of them presented metaplasia, decrease in glandular groups and in turn both disorders, these being considered as important injuries for the development of a gastric carcinoma. The 100% of patients who underwent this procedure showed clinical signs of gastrointestinal disease, from mild stomach diseases until neoplastic infiltrates in advanced stage. It can be concluded that both the presence of nitrite and achlorhydria are fundamental risk factors for the onset of gastric atrophy and metaplasia.

KEY WORDS: achlorhydria, -ATROPHY -ENDOSCOPIA -ENDOSCOPY - ENDOSCOPY HIGH

DIGESTIVE -metaplasia,

INDEX

1. INTRODUCTION

The canton Riobamba is an Ecuadorian subnational territorial entity of the Province of Chimborazo. Its cantonal capital is the city of Riobamba, where most of its total population is grouped, has a population of 263,412 inhabitants and added to the nearby cantons of Colta, Guano and Chambo that are part of "La Y" Metropolitana, has a total population of 365,358 inhabitants (Inen, 2012). The city has several health care centers, both basic and specialized, one of them being the Solca Riobamba Oncology Hospital, which is the site of this study.

For gastric atrophy and metaplasia to occur, the presence of certain factors is indisputable, including the absence of normal stomach acidity, the presence of nitrites or *Helicobacter pylori* colonization. The colonization of the gastric mucosa by the bacterium conditions different inflammatory lesions, in an initial stage it produces an inflammation with an inflammatory infiltrate that extends to the entire depth of the mucosa from the surface and foveolar zone to the formation of lymphoid follicles, which in time produces chronic gastritis.

With inflammation begins a degenerative process of the gastric mucosa, leading to the destruction of the gastric glands (atrophy). The atrophic lesion, initially antral, is not uniform, being observed between preserved areas of the mucosa and can coexist or be replaced in some cases by intestinal type epithelium (intestinal metaplasia), and its subsequent progression to dysplasia, being observed the presence of other advanced forms such as lymphoma of lymphoid tissue associated to the mucosa and gastric carcinoma.

Hypochlorhydria and achlorhydria refer to a decrease in the secretion of hydrochloric acid in the stomach, taking as a normal parameter the pH value of the gastric mucosa is 2.5 (Pérez E.Abdo J. Bernal F 2012). One or the other condition can occur spontaneously as a result of a clinical disorder, or by drug administration (iatrogenic).

Spontaneous hypochlorhydria or achlorhydria may depend on many diseases. Certain surgical procedures can also decrease acid production in the stomach, or eliminate it. The most common cause of spontaneous hypochlorhydria or achlorhydria is active or persistent inflammation of the stomach (chronic atrophic gastritis).

Iatrogenic hypochlorhydria or achlorhydria caused by drug administration may be intermittent or persist throughout the day, depending on the drug taken. Ingestion of antacids or antihistamines rarely elicits continuous decreases in acid secretion throughout the 24-hour period. However, proton pump inhibitors (hydrogen-potassium ATPase inhibitors) may produce persistent hypochlorhydria or achlorhydria in some individuals.

The highly acidic environment in the stomach has a protective function that inhibits bacterial growth. Hypochlorhydria or achlorhydria decreases acidity and may allow bacterial growth in the stomach and upper small intestine (duodenum). Bacterial contamination comes primarily from saliva and food and possibly from the distal part of the intestine.

Not all the determinants in the progression of gastritis to gastric atrophy and intestinal metaplasia have been fully elucidated. In addition to *Helicobacter pylori* infection and its genomic diversity, host susceptibility and immune response have been implicated.

These produce gastric atrophy and intestinal metaplasia, in which the production of hydrochloric acid decreases, an indispensable substance for the defense of the gastric mucosa against the action of foreign germs, causing the presence of nitrites in the gastric medium that provoke the degradation of the mucosa. If there is elevation of pH and there is no presence of nitrites it means that the hypoacidity is temporary, on the contrary if there is elevation of pH and presence of nitrites it indicates that the damage is already an atrophy.

Generally in atrophies there is no presence of *Helicobacter pylori* or it migrates to non-injured sites. If we prove this, this low-cost study can be implemented at a national level, allowing biopsies for pathology to be performed only inpatients with elevated pH and positive nitrites. Other patients would be treated with endoscopic and urease results to investigate *Helicobacter Pylori*, thus reducing the cost significantly.

Based on this premise, the research was undertaken with a total of 105 patients from September 2012 to January 2013 who underwent upper gastrointestinal endoscopy (EDA) in order to determine the discomfort originated in this apparatus. The presence or absence of acidity was determined in the gastric juice, as well as the determination of pH, likewise the gastroenterologist doctor took biopsy samples to make the comparison with the histopathological study, here the morphology is determined by microscopy and it is related to the tissue considered as normal.

Of the 105 samples analyzed, 36 were correlated with increased pH and positive nitrites, as well as with the presence of gastric metaplasia and atrophy; the remaining cases presented pathologies related to mild to severe gastropathies. The purpose of this work was to determine that the hypoacidity and nitrites present are causes of gastric atrophy and metaplasia.

1.2. OBJECTIVES

1.2.1 OBJECTIVE GENERAL

To establish the incidence of hypochlorhydria or achlorhydria and nitrites in the gastric mucosa in patients attended at the Gastroenterology Area of the Solca Riobamba Hospital as a trigger for Gastric Atrophy and Metaplasia.

1.2.2. SPECIFIC OBJECTIVES

1.- To determine nitrites and hydrochloric acid and correlate them with gastric metaplasia and atrophy.
To determine the incidence of gastric metaplasia and atrophy in the patients identifying age, sex, educational level, origin, type of work performed. **3.-** To identify hereditary or acquired risk factors.

1.3. HYPOTHESIS

Achlorhydria and nitrites are triggering causes for the presence of gastric metaplasia and atrophy.

1.4. VARIABLES

Independent: Gastric atrophy and metaplasia.
Dependent: Nitrite test, hypochlorhydria or achlorhydria.
Interveners: Patients with gastric discomfort attending the gastroenterology department.

2. THEORETICAL FRAMEWORK

2.1 ANATOMY AND HISTOLOGY OF THE GASTROINTESTINAL SYSTEM .

The abdomen contains most of the digestive tract, the stomach, small intestine and large intestine, as well as the liver, pancreas and spleen (Rouviere, 2001).

2.1.1 FUNCTIONAL ANATOMY OF THE STOMACH

The stomach is the reservoir where the crushing of foodstuffs initiated in the oral cavity is completed and their digestion begins (Rouviere, 2001).

The stomach is a J-shaped dilatation of the alimentary canal, continuous to the esophagus in the proximal sector and to the duodenum in the distal sector, and functions mainly as a reservoir to store large quantities of freshly ingested food, thus allowing intermittent ingestions, which initiate the digestive process and release their contents in a controlled manner towards the rest of the digestive tract to adapt to the smaller capacity of the duodenum. The volume of the stomach ranges from about 30 mL in a neonate to 1.5 to 2 L in an adult (Sleisenger - Fordtran, 2002).

The viscera are the organs contained within the body cavities and functionally related to the major trophic activities of nutrition: digestion, respiration, circulation, secretion, excretion.
The viscera are covered by conjunctival or epithelial membranes called serosa, such as the peritoneum, pleura and pericardium (Sleisenger-Fordtran, 2002).

The digestive system comprises the set of organs responsible for taking food from the external environment to replace the losses caused by daily cellular work (Sleisenger - Fordtran, 2002).

2.1.2 Embryonic development

The digestive system comes from the inner leaf of the embryo or endoderm. In the first stages, a blind duct called primitive intestine is formed, which extends from its cephalic extremity to its caudal extremity. Upon contact with the ectoderm, the stomodeum or future mouth is formed at the top and the protodeum or future anus, towards the bottom. These orifices are initially obstructed by epithelial walls that are later reabsorbed (Sleisenger - Fordtran, 2002).

The development and rotation of the stomach appears in the fourth week of gestation (28 days), as a dilatation of the distal intestine. As the stomach enlarges, the dorsal side grows faster than the ventral side, to form the greater curvature. Also during the enlargement process it rotates its longitudinal axis about 90 degrees and the greater curvature (the dorsal side) is oriented to the left and the lesser curvature (ventral side) to the right. The combined effects of rotation and growth differences result in the stomach being located transversely in the upper middle and left side of the abdomen (Sleisenger_ Fordtran, 2002).

Embryologically, almost all the organs of the digestive system develop from the endoderm of the primitive intestine (Hib, 1999). The digestive system has an important role in the organism (García-Conde-Merino, 1995), in which the processes of digestion, absorption and excretion of substances necessary for the functioning of other systems of the living being are involved (Guyton, 1999).

The esophagus is relatively wide and dilatable; at its pharyngoesophageal origin it is slightly constricted. This initial narrowing of its lumen is caused by the prominence of the ventral part of the mucosa, under which there is a thick layer of glands. The stomach, the most dilated region of the digestive tract, is a sac-like structure and can be divided into cardia, fundus, body and antrum, pyloric canal and orifice, the fundus being dorsal to the cardiac orifice (Gartner-Hiatt, 1997).

2.2 HISTOLOGY.

The histology of the digestive tract is often described in terms of four broad layers: mucosa, submucosa, external muscular and serosa (or adventitia) (Figure N° 1). These layers are similar throughout their length, whose character and thickness vary with functional needs in different regions (Gartner - Hiatt, 1997; Lesson -Lesson, 1980).

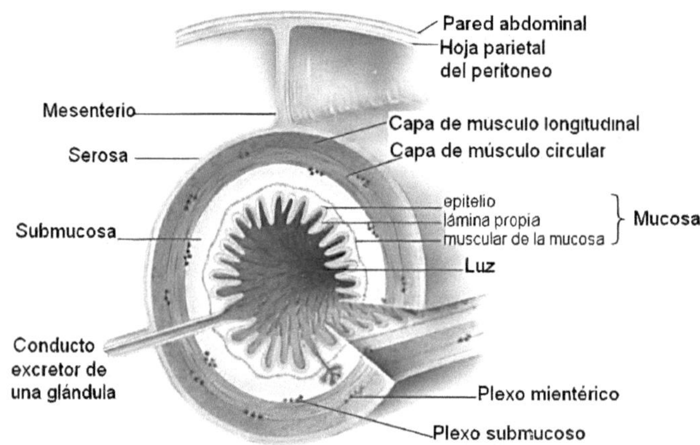

Figure N° 1 Diagram of the Histological layers of the Intestine.

The lumen of the digestive tract is lined by an epithelium, which rests on a deeper layer of lax connective tissue known as lamina propria (Gartner - Hiatt, 1997).

The epithelium, lamina propria (or propria) and muscularis of the mucosa are collectively referred to as the mucosa (Gartner-Hiatt, 1997). This mucosa is an important barrier separating the luminal environment from that of the abdominal cavity.

This layer is surrounded by dense and irregular fibroelastic connective tissue, and is called the submucosa, which also contains blood and lymphatic vessels as well as a parasympathetic nerve plexus, the submucosal plexus of Meissner, which controls the motility of the mucosa as well as the secretory activities of its glands (Gartner - Hiatt, 1997).

The submucosa is lined by a thick muscular layer called the external muscular layer, which is responsible for peristaltic activity, moving the contents of the lumen along the digestive tract (Gartner - Hiatt, 1997).

The external muscular layer is composed of smooth muscle (except in the esophagus) and ɓusually organized in turn into two layers, an internal circular layer and an external longitudinal layer (Gartner-Hiatt, 1997).

There is also a parasympathetic nerve plexus, called Auerbach's myenteric plexus, located in the middle of the two layers that regulates the activity of the external muscular layer which is wrapped by a layer of thin connective tissue and may or may not be surrounded by simple squamous epithelium of the visceral peritoneum called serosa or adventitia (Gartner -Hiatt, 1997).

The gastrointestinal tract receives its parasympathetic innervation from the vagus nerve, except for the descending colon and rectum, which are innervated by the craniosacral nerves (Gartner-Hiatt, 1997).

The luminal surface of the small intestine is modified to increase its mucosal surface to perform the functions, especially digestive absorption and secretion (Gartner-Hiatt, 1997; Lesson-Lesson, 1980).

Three types of modifications have been observed: circular folds (Kerckring's valves), villi and microvilli (Gartner-Hiatt, 1997) (Figure 2).

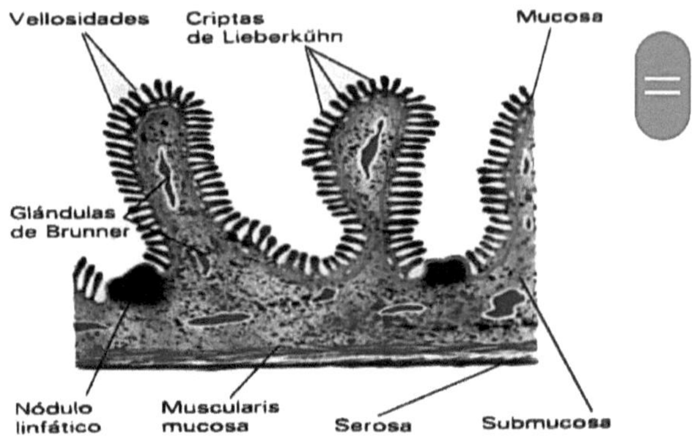

Figure No. 2 Diagram of the small intestine mucosa and its cellular composition.

Circular folds (Kerckring's valves): these are circular or spiral folds of a nucleus that include the total thickness of the mucosa with those of the submucosa (Lesson-Lesson, 1980).

They begin in the duodenum reaching their maximum development in the terminal duodenum, proximal jejunum, and thereafter diminish to disappear in the distal half of the ileum (Gartner-Hiatt, 1997; Lesson-Lesson, 1980).

Villi are small digitiform projections of the mucous membrane, found only in the small intestine, and are covered with epithelium and have a lamina propria nucleus (Gartner-Hiatt, 1997; Lesson-Lesson, 1980).

The center of each villus contains capillary loops, a blind-ended lymphatic duct (chyliferous vessel) and some smooth muscle fibers (Gartner-Hiatt, 1997; Lesson-Lesson, 1980).

The crypts of Lieberkühn are tubular structures that end between the bases of the villi

14

and extend deep into the thickness of the membrane.

mucosa until reaching a point near the mucosal muscle (Lesson-Lesson, 1980).

 The cylindrical absorptive cells lining the villi and lining the crypts have a striated border with innumerable prolongations or microvilli (Lesson-Lesson, 1980).

The lamina propria of the small intestine, which extends to the muscular layer of the mucosa, is compressed into thin sheets of highly vascularized connective tissue because of the numerous tubular intestinal glands, called crypts of Lieberkühn which are composed of superficial absorptive cells, goblet cells, regenerative cells, enteroendocrine cells and Paneth cells which are clearly distinguished by the presence of large apical eosinophilic secretory granules (Gartner-Hiatt, 1997). (Figure N° 2). The ileum has permanent aggregations of lymphoid nodules collectively known as Peyer's plaques (Gartner-Hiatt, 1997).

The submucosa of the duodenum harbors glands, which have been named Brunner's gland, which are branched tubuloalveolar glands whose secretory portions resemble mucosal acini (Gartner-Hiatt, 1997).

The colon lacks folds and villi, thus the surface epithelium is more patent than in the small intestine (Lesson-Lesson, 1980), and is richly endowed with tubular glands called crypts of Lieberkünh, which are similar in composition to those of the small intestine, except that they lack Paneth cells, these tubular glands extend straight from the surface through the entire mucosa to the muscle (Gartner-Hiatt, 1997). (Figure 3).

Gastrointestinal endoscopy known as upper and lower gastrointestinal endoscopy is useful as a complementary examination and if it is indeed reliable to confirm or rule out pathologies that may cause alterations in the digestive system (Gartner-Hiatt, 1997).

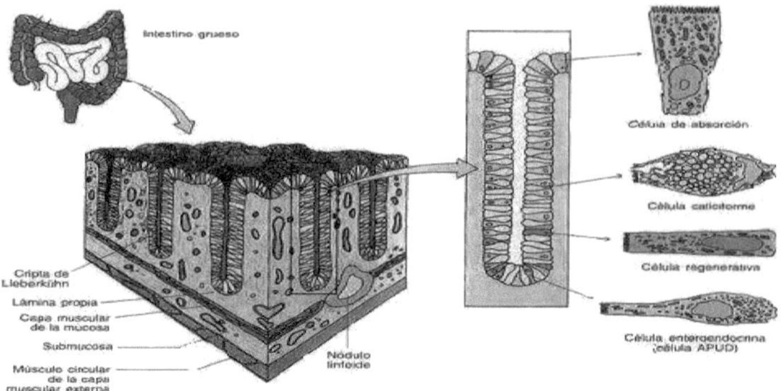

Figure No. 3 Diagram of the large intestine mucosa and its cellular composition.

2.2.1 GASTRIC SECRETION

PHYSIOLOGY

The stomach secretes water, electrolytes, enzymes with activity at acid pH (pepsin and lipase) and glycoproteins (intrinsic factor, mucin). Gastric juice also contains small amounts of calcium and magnesium as well as trace amounts of zinc and iron (Sleisenger-Fordtran, 2002) (Annex 4).

2.2.2 EPITHELIAL CELLS EXOCRINE CELLS

The exocrine cells of the stomach are derived from stem cells located in the mid-region (neck) of the gastric glands. The flow from the neck cells to the surface is a rapid process (<1 week), whereas the downward flow from the neck cells to the gastric glands may require several weeks as the undifferentiated cells mature and transform into more specialized cells, such as parietal cells and chief cells (Sleisenger-Fordtran, 2002).

16

The cylindrical cells that cover the surface of the stomach and its phobias (surface cells) secrete sodium in exchange for hydrogen, carbonates, mucin and phospholipids, all of which help protect the gastric mucosa from injury due to luminal acid pepsin and ingested toxins (Sleisenger-Fordtran, 2002).

2.2.3 GASTRINA

Gastrin is the most potent endogenous stimulant of gastric acid secretion; it is not a single peptide, but belongs to a family of peptides of various lengths, which is obtained from the processing of a larger procurer of 101 amino acids (pre-progastrin), (Sleisenger-Fordtran, 2002).

2.3 PHYSIOLOGY GASTRIC

The stomach plays an important role in human nutrition and has secretory, motor and humoral functions. The stomach has numerous physiological functions:

- It serves as a food store for the time necessary for gastric secretions to act on the food and initiate digestion.
- It produces gastric juice, which contains acid and pepsinogen.
- It mixes the food to reduce the particle size of the food and to give an outlet to the chyme, at the required speed.
- It intervenes in the control of appetite and hunger.
- Controls the bacterial flora that reaches the small intestine avoiding bacterial overgrowth.
- It intervenes in hematopoiesis through the secretion of intrinsic factor which is necessary for the absorption of vitamin b-12 (Drucker, 2005).
- It protects the gastric mucosa from its own peptic acid secretion and duodenal juice by maintaining an intact mucosal barrier.
- It produces and releases substances that act via endocrine or paracrine pathways to regulate metabolic digestive processes (Dvorkin, 2003).

2.3.1 GASTRIC SECRETION

COMPOSITION OF GASTRIC JUICE

Gastric juice is the liquid secreted by the gastric glands into the stomach. It is an aqueous solution containing inorganic components such as chlorine (Cl^-), hydrogen ions ($H+$), potassium ($K+$), sodium ($Na+$) and bicarbonate (HCO_{-3}), and organic components such as mucus, pepsinogens I and II and intrinsic factor. The pH is very low (around 2.5), (Pérez-Abdo-Bernal, 2012).

2.3.2 MEASUREMENT OF THE SECRETION OF ACID

INDICATIONS FOR THE EVALUATION OF SECRETION

Hydrochloric acid is secreted by parietal or oxyntic cells at a concentration of approximately 160 mmoles/L at pH 0.8 (Pérez-Abdo-Bernal, 2012).

The resting parietal cells have unique structures, the vesicular tubule apparatus containing the proton pump ($H+/K+$ ATPase), and the intracellular secretory caniculi, which have a large number of long microvelloci that allow them to increase four to five times the luminal surface area (Pérez-Abdo-Bernal, 2012).

2.3.3 SECRETION OF PEPSINOGEN

Pepsinogen, which is the main proenzyme of gastric juice, is the inactive precursor of pepsin. It is secreted by the principal cells in the glands of the gastric mucosa (Pérez-Abdo-Bernal, 2012).

Pepsinogens are classified into pepsinogen type I and II. The predominant pepsinogen I is secreted by the glandular cells of the mucosa of the gastric fundus or body and is detected in the serum and eliminated in the urine as uropepsinogen. Pepsinogen II is secreted in the fundus, antrum, cardia and proximal duodenum (Pérez-Abdo-Bernal, 2012). In the acidic environment of the stomach pepsinogen is activated to pepsin by cleavage of the N-terminus. This activation occurs only at pH less than 5. At pH 5 to 3 the activation of pepsinogen to pepsin is slow, and at pH less than 3 it is very fast.

Once pepsinogen is activated, its activity depends on pH. Its activity is optimal at pH between 1.8 and 3.5 pH values higher than 3.5 reversibly inactivates pepsin, and pH values higher than 7.2 irreversibly inactivates it (Pérez-Abdo-Bernal, 2012).

Achlorhydria, which results from destruction of parietal cells, occurs in advanced stages and hypochlorhydria can occur even with a large number of preserved parietal cells, suggesting that some anti-proton pump antibodies may be present (Odze-Goldblum, 2009).

2.3.4 OTHER FUNCTIONS

Another enzyme secreted in the stomach is gastric lipase which acts on the fat found in milk and cleaves short-chain triglycerides from fatty acids and monoglycerides (Pérez-Abdo-Bernal, 2012). It works optimally at pH 4 to 7; this enzyme is of limited activity in the stomach of adults; it is of greater importance in children (Pérez-Abdo-Bernal, 2012).

Hypochlorhydria and achlorhydria are a cause of bacterial overgrowth in the proximal intestine and are associated with increased incidence of patients with traveler's diarrhea syndrome (Greenberger-Blumberg-Burakoff, 2009).

2.3.5 SECRETION OF MUCUS

Gastric mucus is a viscous gel made up of 5% glycoproteins and 95% water, with a viscosity of 30 to 269 times that of water; it is adhered to the epithelial cell layer and has a thickness of 5 mm (Schwartz, 2005).
This mucosal gel layer provides protection against injury from harmful luminal

substances, including acid, pepsins, bile acids and ethanol. In addition, it lubricates the gastric mucosa to minimize the abrasive effects of intraluminal food (Latarjet, 2005).

When the mucosa comes into contact with a very low pH solution, mucus precipitates and mucous cells must constantly secrete mucus (Latarjet, 2005).

Measurement of gastric acid secretion can aid in the clinical diagnosis and management of patients with gastrinoma and other hypersecretory acid states, in the diagnosis of incomplete vagotomy in patients with recurrent post-surgical peptic ulcer (Sleisenger-Fordtran, 2002).

Likewise, the demonstration of fasting acid (or fasting acidic gastric pH) excludes achlorhydria as a cause of a markedly elevated serum gastrin concentration. Patients should discontinue gastric antisecretory drugs before measuring fasting acid secretion (Sleisenger-Fordtran, 2002).

2.3.6 BASAL ACID PRODUCTION (PBA)

PBA represents gastric acid secreted without intentional or avoidable stimulation. About 2 to 3 normal people secrete a certain amount of gastric acid under basal conditions, the upper limit of normal ABP is about 10 mmol per hour in men and 5 mmol per hour in women (Dalenback, 1996).

PBA fluctuates from hour to hour in the same person. The lowest ABP occurs between 5 and 11 am and the highest between 2 pm and 11 pm. The variation in ABP also depends on cyclic gastric motor activity, probably due to fluctuations in cholinergic tone (Sleisenger-Fordtran, 2002).

2.3.7 ACID SECRETION STIMULATED BY FOOD

Gastric acid secretion rates after ingestion increase rapidly and approach the value of peak acid production (PPA). Despite this, the pH in the stomach increases because the proteins in the meal neutralize the secreted acid (Sleisenger-Fordtran, 2002).

Then, postprandial intragastric pH decreases below basal pH as gastric acid secretion maintains secretion and buffers act or food leaves the stomach (Sleisenger-Fordtran, 2002).

2.4 PATHOLOGIES THAT AFFECT THE GASTROINTESTINAL SYSTEM AND CAN BE DIAGNOSED BY PERFORMING AN ENDOSCOPIC EXAMINATION WITH SUBSEQUENT TAKING OF BIOPSY.

The gastrointestinal system is affected by a large number of pathologies, however this chapter refers to pathologies that can be diagnosed by endoscopy and/or biopsy (Yazbeck, 1995).

2.4.1 ANOMALIES OF THE DEVELOPMENT

Congenital esophageal anomalies are relatively common (1 in 3000 to 1 in 4500 live births), and are due to genetic defects or intrauterine stress that impede fetal maturation (Yazbeck, 1995). Esophageal anomalies are common in preterm infants, and 50% have other disorders, which are summarized by the acronym VACTERL (formerly VATER), (Yazbeck, 1995).

2.4.2 ESOPHAGEAL ATRESIA AND TRACHEOESOPHAGEAL FISTULA

Esophageal atresia and gastroesophageal fistulas are the most important and most common developmental esophageal anomalies. The first results from the defect of recanalization in the anterior primitive intestine; the others arise from failure of the pulmonary bud to separate completely from the anterior intestine (Spitz, 1996). Esophageal atresia occurs as an isolated anomaly in only 7% of cases; the remaining 93% are accompanied by one of the forms of tracheoesophageal fistula (Spitz, 1996).

In isolated atresia the upper esophagus ends in a blind pouch and the lower esophagus connects to the stomach. This pathology can be suspected before birth by the development of polyhydramnios (due to the inability of the fetus to swallow and thus absorb amniotic fluid) or at birth by regurgitation of saliva and an excavated (airless) abdomen (Pope, 1993).

2.4.3 CONGENITAL STENOSIS

Congenital stenosis is a rare anomaly occurring in only 1 in 25,000 live births. The stenosed segment varies from 2 to 20cm in length and is usually located within the middle or lower third of the stomach. The precise cause of congenital stenosis is not fully understood (Sleisenger-Fordtran, 2002).

When much of the stenosed walls are resected, they contain sequestrations of respiratory tissue (hyaline, cartilaginous, respiratory epithelium), suggesting that their origin is incomplete separation of the pulmonary bud from the anterior primitive intestine (Sleisenger-Fordtran, 2002).

In other cases, stenosis results from fibromuscular hypertrophy or damage to the myenteric plexus with loss of neural elements producing smooth muscle relaxing nitric oxide (Sleisenger-Fordtran, 2002).

2.4.4 ESOPHAGEAL DUPLICATIONS

Congenital duplications of the esophagus occur in 1 in 8000 live births. They arise as epithelium-lined cases from the anterior primitive intestine and develop to produce tubular or cystic structures that do not communicate with the esophageal lumen (Sleisenger-Fordtran, 2002). Cysts make up 80% of duplications and are usually single fluid-filled structures (Sleisenger-Fordtran, 2002).

2.4.5 ESOPHAGEAL RINGS

The distal esophagus contains two "rings", a and b, which anatomically demarcate the proximal and distal borders of the esophageal vestibule (Chotiprasidhi, 2000).

2.4.6 PATHOLOGIES AFFECTING THE STOMACH AND INTESTINE THAT ARE DIAGNOSED BY ENDOSCOPY AND/OR BIOPSY:

The term gastritis, which means inflammation of the stomach, is one of the most heterogeneously interpreted medical concepts, since it constitutes a diverse and multicausal pathological process (Ramos, 2000).

This term refers to a series of entities where there is damage of the gastric mucosa with presence of inflammatory infiltrate; it is usually caused by infectious agents, hypersensitivity, autoimmune or idiopathic reactions and therefore, its diagnosis is established only and exclusively with histology by biopsy (Ramos, 2000). The term gastropathy is used when there is gastric mucosal damage in which the inflammatory infiltrate is minimal or absent, and the predominant alteration is epithelial (reactive gastropathy) or vascular (congestive, ischemic, etc.), (Ramos, 2000).

2.4.6.1 CLASSIFICATION

By their macroscopic appearance and etiology they are classified as specific and nonspecific; by the type of inflammatory cells they are divided into acute and chronic; by their location in type A when it covers the gastric fundus and body (usually related to immunological factors); type B is located in the antrum (mainly associated with H pylori infection) and type AB is a pangastritis (Pérez-Abdo-Bernal, 2012).

ACUTE GASTROPATHIES

Acute gastropathies, also called erosive hemorrhagic gastropathies, the inflammatory infiltrate is minimal or absent, so the more appropriate term is gastropathy, rather than GASTRITIS (Sleisenger-Fordtran, 2002).

It is characterized by the existence of erosions (loss of continuity of the mucosa, without involving the *muscularis mucosae)* or hemorrhagic foci in the mucosa of the stomach or duodenum, which may be few or multiple (Sleisenger-Fordtran, 2002).

Lesions are observed by endoscopy and biopsies are not usually required unless a special type of gastritis is suspected (Sleisenger-Fordtran, 2002).

GASTROPATHIES DUE TO NON-STEROIDAL ANTI-INFLAMMATORY DRUGS

Nonsteroidal anti-inflammatory drugs form a group of drugs of great importance and very frequent use in routine medical practice due to their broad therapeutic spectrum, which encompasses not only their well-known anti-inflammatory, analgesic and antipyretic effects, but their range has expanded in recent years as their efficacy has been demonstrated in many other pathological situations (Hawkey-Langman, 2003).

The local effect, during the passage of the drug through the stomach, involves the gastric pH and the acid structure of the Aines (especially with acetylsalicylic acid), which decreases the protective hydrophobic mucosal barrier which favors the retrodiffusion of hydrogenions and in case of concomitant biliary reflux favors mucosal damage, in addition to the direct participation of active metabolites in the damage, especially with aspirin (Hawkey-Langman, 2003).

GASTRITIS CHRONIC

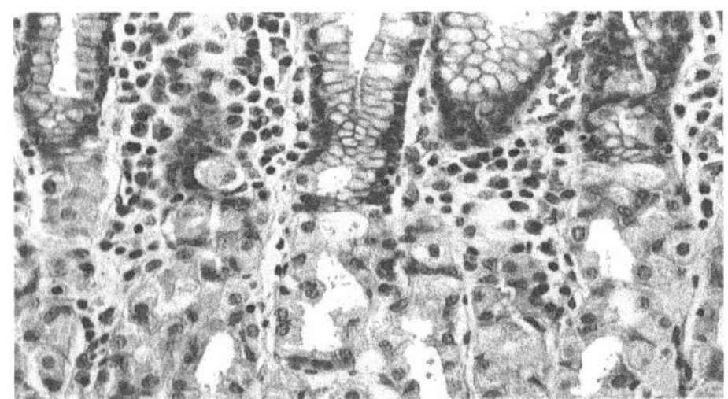

Figure No. 4 Superficial chronic gastritis. H-E staining Source:
http://www.conganat.org/7congreso/trabajo.asp?id_trabajo=521&tipo=1

The term chronic gastritis implies the existence of an inflammatory infiltrate of the gastric mucosa (Zhang, 2005). The momentous discovery that H pylori is involved in the etiology of most cases rethinks previously established concepts (Zhang, 2005).

Chronic erosive gastritis - Erosive gastritis is a form of gastritis in which ulceration occurs in the deepest layer of the stomach lining. This type of gastritis affects people of different ages, being more common in men than in women (Zhang, 2005).

It is closely related to frequent use of NSAIDs (non-steroidal anti-inflammatory drugs), alcohol, presence of Helicobacter Pylori, smoking, etc. These erosions can be shallow to deep and are usually circular in shape and may bleed, leaving the patient at risk of developing anemia or perforation (Zhang, 2005).

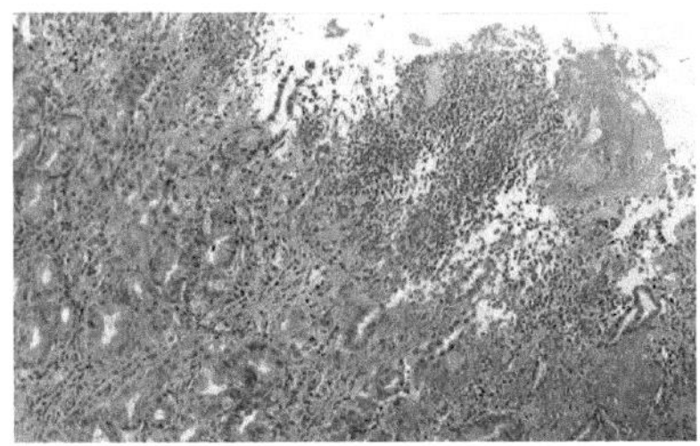

Figure No. 5 Chronic erosive gastritis. H-E staining Source:
http://www.conganat.org/7congreso/trabajo.asp?id_trabajo=521&tipo=1

Chronic non-atrophic gastritis: Histologically, there is an inflammatory infiltrate without destruction or loss of gastric glands (Zhang, 2005).

Superficial chronic gastritis - It has a banded lymphoplasmacytic inflammatory infiltrate, which occupies the superficial portion of the gastric mucosa in foveolae and glandular necks. It is considered that it is not a pathology as such but represents the initial stage of other forms of chronic gastritis associated with alcohol intake, certain spicy foods, medications and infection by H. pylori (Zhang, 2005).

Diffuse antral gastritis - It is characterized by a dense lymphoplasmacytic inflammation that covers the entire thickness of the mucosa of the antrum, expands the lamina propria and separates the gastric glands, giving the false appearance of glandular loss and atrophy, there may be prominent lymphoid follicles that receive the name of follicular gastritis. This type of gastritis is found in almost all cases of duodenal or pyloric ulcer, and the main etiological agent is H pylori (Zhang, 2005).

Chemical or reflux gastritis. - It is considered a special clinical pathological subtype within non-atrophic gastritis. It occurs in patients with post gastrectomy anastomosis or who have persistent duodeno-gastric reflux caused by the presence of bile salts which, combined with isolecithin and pancreatic enzymes, damage the gastric mucosa. Histologically there is stromal edema, expansion and tortuosity of foveolar spaces, vascular congestion, scarce inflammatory infiltrate (Zhang, 2005).

GASTRITIS ATROPHIC

This group consists of two nosologically distinct entities, in which there is reduction and loss of gastric glands (Zhang, 2005).

Diffuse corporal atrophic gastritis - characterized by the loss (atrophy) of the oxyntic glands of the body and gastric fundus, with main affection of the main and parietal cells (acid and intrinsic factor producers) (Zhang, 2005).

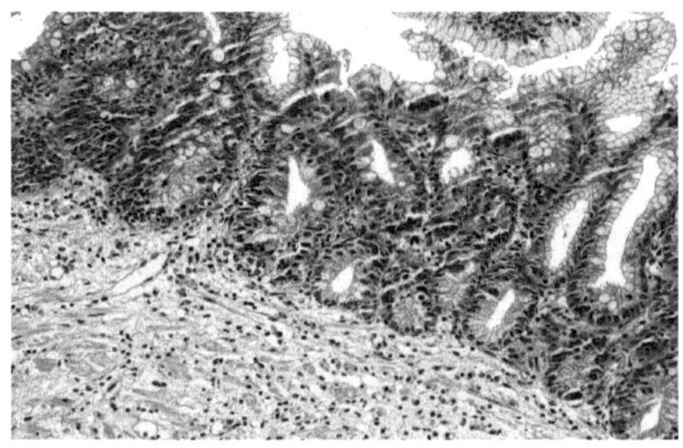

Figure No. 6 Chronic atrophic gastritis. H-E staining Source:
http://www.conganat.org/7congreso/trabajo.asp?id_trabajo=521&tipo=1

It is progressive leading to severe epithelial atrophy with extensive intestinal metaplasia and has a high risk of developing malignant neoplastic lesions originating in the metaplastic territory, with minimal association with gastric ulcers (Zhang, 2005).

Multifocal atrophic chronic gastritis - It is distributed in all continents and racial types, and its presence coincides with the geographic distribution of populations at high risk of gastric ulcer and gastric cancer. Histologically it shows independent foci, with patchy distribution of glandular atrophy and presence of metaplasia which can be of mature phenotype, also called type I, complete metaplasia or small intestine, or immature phenotype, called incomplete metaplasia or large intestine (Jm-Blasco, 2005).

Follicular chronic gastritis - The main characteristic of follicular chronic gastritis is the inflammatory reaction that defines an infiltrate of mononuclear cells, of which monocytes are the main ones, and at the same time, it produces the formation of lymphoid follicles that have a germinal center (De Weerth-Gocht-Seewald-Brand, 2002).

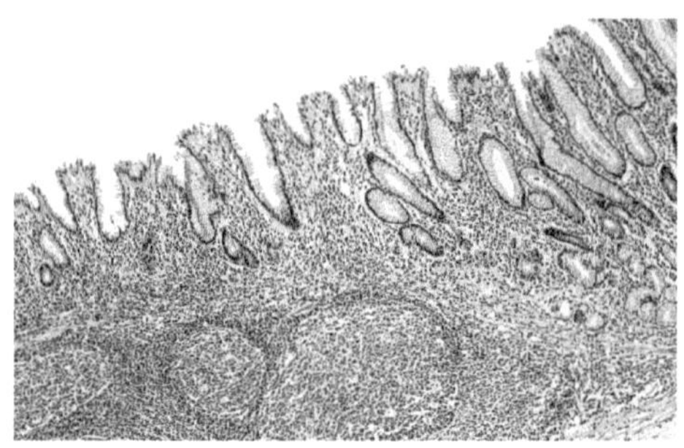

Figure No.7 Chronic atrophic gastritis. H-E staining Source:
http://www.conganat.org/7congreso/trabajo.asp?id_trabajo=521&tipo=1

Eosinophilic gastroenteritis is a rare and poorly understood disorder characterized by diffuse or segmental infiltration of some part of the alimentary canal with mature eosinophils, usually accompanied by peripheral eosinophilia. The disease affects one or more layers of the stomach, small intestine or colon with the resulting clinical syndromes of chronic vomiting (eosinophilic gastroenteritis), (Sainz and Rodriguez, 2004), chronic small bowel diarrhea (eosinophilic enteritis), chronic large bowel diarrhea (eosinophilic colitis) or any combination of these (Wolfe, 2002).

2.5 NEOPLASMS AFFECTING THE GASTROINTESTINAL SYSTEM .

In 1983 Marshall and Warren reported to the scientific community the finding in the stomach of patients with gastritis and peptic ulcer of a Gram-negative spiralized bacterium which they named *Campylobacter like* organism and which today is known as *Helicobacter pylori* (Ramirez-Ramos-Gilman, 2004).

This communication was received with great skepticism because until then it was stated that "no microorganisms could survive in the stomach, due to the acid gastric pH, there being only the possibility that there were passing germs and that the microorganisms described by these Australian authors were due to contamination", (Ramirez-Ramos-Gilman, 2004).

Almost a decade of disbelief and controversy passed until the evidence accumulated during that time on the pathogenic role of this bacterium within the multifactorial nature of peptic, gastric and duodenal ulcers, motivated the World Congress of Gastroenterology, held in Australia in 1990, to recommend "the eradication of *Helicobacter pylori* in all patients with gastric or duodenal ulcers in whom its presence was demonstrated" (Ramirez-Ramos-Gilman, 2004).

Since then, there has been an explosion of communications on research results in various fields: microbiology, molecular biology, epidemiology, mechanisms of pathogenicity, diagnostic methods, treatment schemes, recurrence, reinfection and vaccination (Ramirez-Ramos-Gilman, 2004).

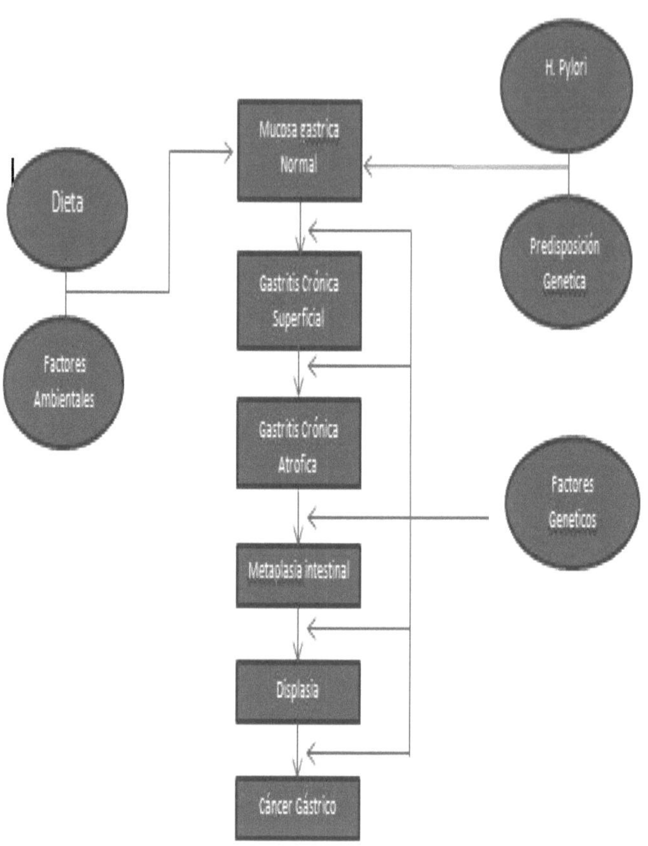

Figure No. 8. Hypothesis of the genesis of gastric cancer associated with infection
Source: Farreras-Rozman, 2000.

On the other hand, in view of the heterogeneity of research results in each of these fields, and therefore the resulting controversies, consensuses were initiated to unify criteria on the pathogenic action, diagnostic methods, treatment schemes, among other aspects: Consensus of the National Institutes of Health of the United States of America (1994), Latin American Consensus (1999), Maastricht Consensus (1997, 1998 and 1999), Consensus of the American College of Gastroenterology (2002), (Ramirez-Ramos-

30

Gilman, 2004).

Thus, it is now accepted that infection by this bacterium plays an important role in the genesis of gastritis, duodenal peptic ulcer, gastric peptic ulcer, gastric cancer and MALT-type lymphoma. Regarding the role that *Helicobacter pylori* may play in the multifactorial etiopathogenesis of gastric cancer, many "epidemiological evidences" have been published, and several hypotheses have been postulated to explain them (Ramírez-Ramos-Gilman, 2004).

On the other hand, several so-called "enigmas" or geographical variations have been exposed. It is considered that from an epidemiological point of view it is not difficult to establish statistically supported conclusions. What is generally difficult is to satisfactorily explain these relationships and this is also the case of the *Helicobacter pylori* - gastric cancer - epidemiology link (Ramirez-Ramos-Gilman, 2004).

2.6 INTESTINAL METAPLASIA

Intestinal metaplasia is a frequent process, especially in areas with a high incidence of gastric cancer.

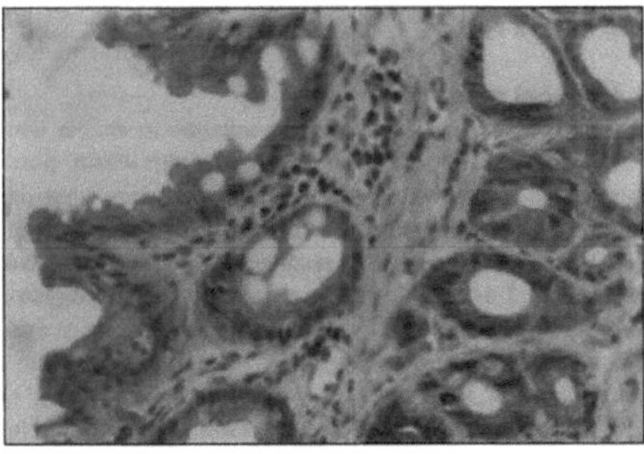

Figure No. 9. Complete intestinal metaplasia. H-E staining Source:
http://www.conganat.org/7congreso/trabajo.asp?id_trabajo=521&tipo=1

It has been estimated that 10-20 years after the diagnosis of intestinal metaplasia, only

10% of patients will develop adenocarcinoma of the stomach. There is a frequent coexistence between intestinal metaplasia, particularly type III, and gastric carcinoma, especially of the glandular type (Correa-Coello-Duque 1999).

Intestinal metaplasia was divided into three groups: type I or complete which has two groups: type IA of thin intestinal mucosa and type IB of colonic mucosa; type II or incomplete which in turn is subdivided into two groups; type IIA (mucoprotein-gastric mucosa) and type IIB (sialomucin-thin intestinal mucosa); and type III also incomplete (sulfomucin-colonic mucosa). Extensive metaplasia has a higher risk of malignization and may be reversible, as has been observed after eradication of H. pylori (Correa-Coello-Duque 1999).

2.7 NITRATES AND NITRITES

In addition to providing us with preformed NA, the diet can be the entry route for nitrates and nitrites which are substrates for the endogenous formation of NA. Vegetables provide approximately 80% of the daily intake of nitrates, the highest concentration being found in beets, celery, radish, chard and spinach (Walters, 1992).

Although there is variability depending on the type of fertilizers used in its cultivation. Drinking well water can be an important source depending on the nitrate content of the soil. In some countries it has been estimated that 10% of nitrate intake came from water (Sugimura, 2000).

Nitrites have two main sources: 40% comes from fresh or preserved vegetables, and the rest from cured and preserved meat and cereals (Cornee-Lairon-Velema, 1992). Nitrites have bactericidal properties and sodium nitrite is used as a preservative in the prevention of botulism, especially during the curing process of meat (Walters, 1992). In some countries the average intake has been estimated at 1.9 mg per person per day (Cornee-Lairon-Velema, 1992).

So far there are no studies estimating the level of population dietary exposure. Nitrate (NO3 -) is reduced to nitrite (NO2⁻) by bacteria in saliva and also in the stomach (Ralt-Tannenbaum, 1981). Approximately 25% of the ingested nitrate recirculates in the saliva, and 20% of it is converted to nitrite. Of the nitrite that reaches the stomach, 20% comes from food, while 80% comes from nitrate reduction in saliva (Suzuki-Iljima, 2003).

The formation of nitrites from nitrates takes place in the stomach and increases with increasing pH, which occurs as a consequence of chronic *Helicobacter pylori* infection. Under specific circumstances, such as chronic gastritis, nitrites can be oxidized in the stomach to nitrosating agents (N2O3, N2O4) and react with secondary amines to form N-nitrosocompounds. Formation Endogenous introduction accounts for 45% to 75% of total NOC exposure (Suzuki-Iljima, 2003).

Ascorbic acid (AA) is a water-soluble vitamin and a potent reducing agent. AA is secreted into gastric juice and plays an important role in preventing nitrosation. It does this by competition with secondary amines through acidification of nitrite. In this reaction the acidified nitrite is reduced to ON (nitric oxide) and AA is oxidized to dehydroascorbic acid. There are two mechanisms involving ON and AA in nitrosation (Suzuki-Iljima, 2003).

The first involves the production of high concentrations of ON in the light from the reaction between nitrite from saliva and AA from gastric juice. This ON diffuses into the epithelium and cells forming nitrosating species which can directly damage DNA (Wink-Fellisch, 1999). The second mechanism involves the generation of nitrosating species within the lumen due to acidification of nitrite by a low concentration of AA (Suzuki-Iljima, 2003).

These nitrosating species react with nitrogenous compounds to form NOCs. The first mechanism is promoted by the presence of AA, while the second by its absence.

Everything depends on the conditions of the medium present at the time. Thus the pH changes that occur in inflammatory processes may influence one mechanism to prevail over the other. Both mechanisms can operate at the same site but not simultaneously (Suzuki-Iljima, 2003).

It is also said that the diet can introduce nitrites and N-nitroso compounds that can produce hypochlorhydria due to atrophy of parietal cells with the consequent bacterial overgrowth that added to metaplasia increase the risk of gastric cancer (Suzuki-Iljima, 2003).

The detection of high levels of nitrite saliveres swallowed is rapidly converted to nitric oxide and an acid-catalyzed chemical reaction takes place at the gastroesophageal junction leading to inflammation, metaplasia and subsequently neoplasia (Lijima- Shimosegawa, 2006).

Dietary and environmental factors

There is a strong relationship between the incidence of gastric cancer and a diet with a high salt intake and low in fresh fruits and vegetables, low in vitamins A, C and E and micronutrients (selenium), as well as preservation methods with a possible carcinogenic effect, such as smoking, salting and pickling (Correa-Coello-Duque, 1999).

Gastric cancer has also been linked to the concentration of nitrites in the diet and drinking water. Bacteria present in the mouth and stomach would reduce nitrites to nitrates that could lead to the formation of nitrosamides and nitrosamines, known to have mutagenic and oncogenic effects. Gastric hypoacidity, vitamin C and E deficiency and bacterial contamination of low-quality food consumed by the poorest strata of the population would act in favor of this mechanism (Correa-Coello-Duque, 1999).

The danger of nitrate, a substance that is not toxic in itself, lies in its chemical transformation into nitrite, which occurs, in part, during human metabolism. This nitrite can react in the acid medium of the stomach with amines, substances obtained from the metabolism of protein foods (meat, fish, eggs, milk and derivatives of these foods), giving rise to nitrosamines, which are carcinogenic agents (Correa-Coello-Duque, 1999).

The acid pH of the stomach favors the formation of N-nitrosamines from nitrites and secondary amines, since it has been observed that nitrites, in the presence of acidic conditions in the stomach, carry out nitrosation. Some bacteria are also producers of nitratoreductase enzymes, capable of reducing considerable amounts of nitrate to N-nitrosamines (Correa-Coello-Duque, 1999).

Such bacterial catalysis can be triggered by denitrifying bacteria such as *Pseudomonas aeruginosa*, *Neisseria* and non-denitrifying bacteria such as *Escherichia coli* capable of increasing the biosynthesis of N-nitroso compounds in the stomach (Correa-Coello-Duque, 1999).

Other environmental factors such as alcohol and tobacco consumption are not well related to the development of gastric cancer. Nor has the development of adenocarcinoma of the stomach been demonstrated after prolonged administration of antisecretory medication such as H2-antagonists or proton pump inhibitors (Correa-Coello-Duque, 1999).

Ascorbic acid can block this nitrosation reaction. In relation to this action of ascorbic acid, a decrease in the level of ascorbic acid in gastric juice has been observed in chronic gastritis with elevated pH and Helicobacter pylori infection (Correa-Coello-Duque, 1999).

Likewise, patients with intestinal metaplasia have low serum levels of ascorbic acid compared to healthy patients. On the other hand, it has been observed that ascorbic acid intake is associated with a decreased risk of gastric cancer (Correa-Coello-Duque, 1999).

The level of individual exposure to nitrosamines (NA) depends on an individual's diet, lifestyle and occupation. Exposure can occur exogenously, through ingestion of preformed NAs that are present in food, tobacco use and/or occupational exposure or endogenously, where NAs are synthesized in the body from precursors from the diet (Han-Peura, 2008).

2.8 INTRODUCTION TO GASTROINTESTINAL ENDOSCOPY HIGH

Digestive endoscopy as a method of diagnosis and treatment of digestive diseases has existed for the last one hundred and twenty years. Although the first endoscopes were described in 1868, the new era of digestive endoscopy emerged in 1957 with the appearance of fiberoptic endoscopes (L-Abreu, 2006).

This technological advance has meant a positive revolution in the management of patients with digestive diseases. In little more than forty years, endoscopy has been developing in such a way that medical treatises have had to be revised to update the diagnosis and treatment of many diseases (L-Abreu, 2006).

2.8.1 ENDOSCOPIC ANATOMY OF THE STOMACH

The endoscopic orientation in the stomach is as follows: The lesser curvature is located at 12 o'clock; the greater curvature at 6 o'clock; the anterior wall, at 9 o'clock; and the posterior wall at 3 o'clock. The regions of the gastric cavity are described below (Naylor-Axon, 2003).

a) **Fundus.** It is the area located above a line drawn transversally to the cardia; it is explored with endoscopy in retroflexion, appearing as an area in continuity with the cardia and anterior to it, with a convex morphology and flat mucosa without folds (L-Abreu, 2006).

The submucosal vascular pattern is clearly visualized without implying mucosal atrophy (unlike in the gastric body), being composed of capillaries and even fundic veins that are straight and not very prominent (which differentiates them from the fundic varices of portal hypertension), (L-Abreu, 2006).

In the most distal portion of the fundus at its junction with the greater curvature, gastric secretions accumulate giving rise to the so-called gastric cascade or mucosal lake. The presence of bile in the gastric cascade is constant in operated stomachs and is very frequent in cholecystectomized patients, but in the rest of the cases it is of unspecific clinical significance (L-Abreu, 2006).

b) **Gastric body.** This is limited by the fundus in the proximal part and by a line drawn transversely through the angular incisura in its distal part. This lower limit usually corresponds to where the rough mucosa changes to flat mucosa. The main characteristic of the gastric body is the presence of gastric folds, more prominent in the greater than in the lesser curvature (L-Abreu, 2006).

c) **Angular incisura.** This is located in the lesser curvature and separates the gastric body from the antrum. It is best visualized in retroflexion and appears as an irregular symmetrical fold of 5 to 10 mm with a smooth surface. It is important its correct evaluation since it is a frequent seat of ulcer pathology (L-Abreu, 2006).

d) **Antrum.** It presents an important morphological variety. The antrum can be short (<3 cm) or long (>10cm), it can have an axis parallel to that of the gastric body or deviate anteriorly.
or later, prominent antral folds, which are nonspecific in themselves, may be observed in 10% of patients (L-Abreu, 2006).

2.8.2 MAINCAUSES FOR FOR ENDOSCOPY UPPER GASTROINTESTINAL ENDOSCOPY

DYSPEPSIA

It is a very frequent symptom, which is the main reason for requesting an oral endoscopy. Due to the different interpretations of this term and in order to avoid confusion, in 1997 it was defined in Maastricht as pain or discomfort located in the upper abdomen, including nausea and vomiting, early satiety, regurgitations or epigastric swelling, but not heartburn or dysphagia (García-Conde-Merino, 1995).

The prevalence of dyspeptic symptoms in the general population varies between 14% and 41%, with large geographical variations (Garcia-Conde-Merino, 1995).

GASTROESOPHAGEAL REFLUX DISEASE (ERGE)

It consists of the existence of esophageal histological alterations in patients with reflux symptoms. The main symptom of gastroesophageal reflux is heartburn or retrosternal burning sensation, which occurs at least once a month in 40% of the general population. The prevalence of esophagitis in these patients has been estimated at 3-7 % with an incidence of 120/100,000 inhab/year (García-Conde-Merino, 1995).

BARRETT'S ESOPHAGUS

It consists of a substitution of the squamous epithelium of the distal esophagus along 3 or more centimeters by metaplastic columnar epithelium, as a consequence of a gastroesophageal reflux of long evolution. These patients have a higher risk of developing esophageal adenocarcinoma than the general population (García-Conde-Merino, 1995).

ATYPICAL CHEST PAIN

It is defined as the existence of pain at the thoracic level without evidence of coronary involvement. The exact mechanism of pain is not known, but it has been postulated that chemo-, thermo- or mechanoreceptors would be stimulated by the action of acid or pepsin, distension or temperature, respectively.

The main causes of pain are gastroesophageal reflux and motility disorders, but it is important to emphasize that a normal endoscopy does not rule out an esophageal cause of chest pain (García-Conde-Merino, 1995).

DYSPHAGIA AND ODYNOPHAGIA

Dysphagia consists of difficulty or delay in swallowing, distinguishing two forms; oropharyngeal dysphagia, which denotes difficulty in transferring the bolus from the oropharynx to the upper esophagus and is usually due to a neurological cause, and esophageal dysphagia secondary to disorders in the esophageal body. Odynophagia or pain when swallowing suggests mucosal inflammation or spasm, being its main causes: caustic ingestion, esophagitis of any cause and infections (candida, herpes or cytomegalovirus), (García-Conde-Merino, 1995).

UPPER GASTROINTESTINAL BLEEDING

Upper gastrointestinal bleeding (UGH), manifested by hematemesis, melena or hematochezia, is a very common problem. The main cause is peptic ulcer followed by esophageal varices and neoplasms (L-Abreu, 2006).

The clinical signs of patients with inflammatory bowel disease vary according to the severity and location of the cellular infiltration. Patients in whom the small intestine and gastric mucosa are affected generally present chronic vomiting, weight loss and diarrhea, while those in whom the large intestine mucosa is affected may present chronic tenesmus, frequent defecation, hematochezia and mucus in stool (Rojo, 2003).

DEFINITION OF KEY WORDS

CHLORHYDRIA. Lack of hydrochloric acid in gastric secretions.

ATROPHY. It is the decrease in cell size by loss of cellular substance **ENDOSCOPY.**
It is a diagnostic and therapeutic technique, used mainly in medicine, which involves
the introduction of a camera or lens inside a tube or endoscope through a natural
orifice, a surgical incision, a lesion for visualization of a hollow organ.

**ENDOSCOPYADIGESTIVEENDOSCOPYADDIAGNOSTICTH
ERAPEUTICADDITION   .**

Visualizes the esophagus, stomach and duodenum. It detects cancers of these regions, it
is the study of choice in upper gastrointestinal bleeding, among other utilities.

METAPLASIA. A reversible change in which an adult-type cell (epithelial or
mesenchymal) is replaced by another adult-type cell.

3.- MATERIALS AND METHODS

3.1. MATERIALS

3.1.1. LOCATION OF THE RESEARCH

The study was carried out at the **SOLCA CHIMBORAZO Oncology Hospital** in the city of Riobamba.

3.1.2. PERIOD OF THE RESEARCH

The period of the investigation was from September 2012 to January 2013.

3.1.3. RESOURCES EMPLOYEES

3.1.3.1. Resources Human Resources

- The researcher.
- Tutor
- Medical specialist in Gastroenterology.
- Nursing Assistant
- Medical specialist in Anatomical - pathology

3.1.3.2. Resources Physical.

- Potable alcohol
- Pens.
- Coloring H-E Technique
- Laptop Computer
- Disposable blades (*Leyca* brand)
- Dormicum - Midazolam (Patient Sedation)

- Entellan (Plate Mounting Resin)
- Eosin
- Absolute ethanol
- 10% buffered formalin
- Gelatin p.a
- Stapler.
- Gloves
- Hematoxylin
- Printer.
- Solca Riobamba Pathology Laboratory.
- Masks
- Filter paper
- Cover plates 24x40mm
- Object carrier plates
- Ream of bond paper.
- Ink for the printer.
- Nitrite strips (COMBUR TEST)
- pH strips
- 250 mL beakers
- Xylene

3.1.3.3 Equipment

- Filing cabinets for kerosene blocks and slides
- Flotation bath
- Bain Marie
- Staining cuvettes
- Kerosene dispenser
- Stove

- Gastroscope
- Microscope
- Microtome
- Refrigerator, freezer
- Heating plates
- Tissue processor

3.1.4. UNIVERSE

The universe consisted of all patients admitted to the gastroenterology area of the **SOLCA RIOBAMBA Hospital** during the research period September 2012-January 2013. (105 patients).

3.1.5. SAMPLE

The sample consisted of all patients with hypochlorhydria and nitrites in the gastric mucosa (36 patients).

3.2. METHODS

3.2.1. TYPE OF RESEARCH

- Descriptive
- Experimental

3.2.2. DESIGN OF RESEARCH

The research design was non-experimental.

Sample collection protocol

- The patient must be fasting for at least 8 hours prior to the test.

- The nursing assistant is in charge of taking vital signs (blood pressure, temperature, pulse).

- The physician requests that the patient be given the recommended dose of dormicum (midazolam) for pseudoanalgesia.

- Once the patient is sedated, the physician introduces the gastroscope, assesses the esophagus, larynx and reaches the stomach (body, fundus).

- Here by the introduction of the gastroscope, the patient begins to eliminate reflux and the sample is taken in a sterile container to detect the presence of nitrites and to assess the pH.

- After this, the doctor takes a sample of the lesion (biopsy), the most representative part.

- The biopsy is fixed in buffered formalin (24 hours) and then enters the processing battery.

- After this, the biopsy is prepared (histological section), stained and the plate is ready for diagnosis by the anatomist-pathologist.

- The results were recorded on a daily production sheet in order to keep a monthly control and create a database (Annex 1).

- Patients positive for achlorhydria and nitrites were followed up with histopathological study of the biopsy and the results were correlated for their respective tabulation.

4. RESULTS AND DISCUSSION

PERCENTAGE OF PATIENTS CLASSIFIED BY SEX WHO UNDERWENT UPPER GASTROINTESTINAL ENDOSCOPY. SOLCA CHIMBORAZO. SEPTEMBER 2012 - JANUARY 2013.

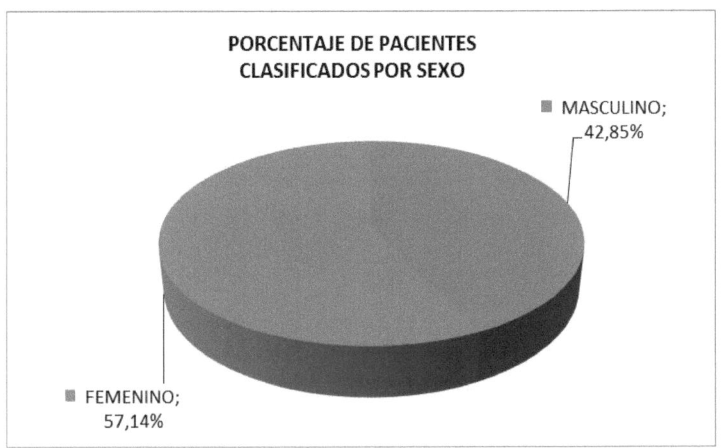

Graph No. 1. Patients classified by sex who underwent upper gastrointestinal endoscopy. Solca-Chimborazo between September 2012 and January 2013.

Of a total of 105 patients attended during this study, 60 corresponded to the female sex with a total of 57.14 %, and 45 to the male sex with 42.85 %, all of them presented discomfort in the digestive tract.

PERCENTAGE OF PATIENTS BY AGE INTERVALS WHO UNDERWENT UPPER GASTROINTESTINAL ENDOSCOPY. SOLCA CHIMBORAZO. SEPTEMBER 2012 - JANUARY 2013.

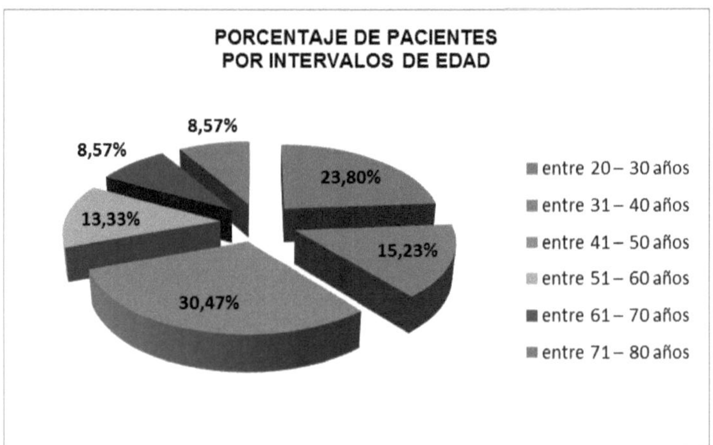

Graph No. 2.   **Patients classified by age range who underwent upper gastrointestinal endoscopy. Solca-Chimborazo between September 2012 and January 2013.**

Of the majority of patients analyzed in this study, the group between 41 to 50 years of age stands out with a total of 30.47%, then the group between 20 to 30 years with a total of 15.23%, and the lowest incidence between 61 to 80 years with 8.57%, so it can be said that in any age group the lesions with gastric discomfort are present, what should be determined is the predominant factor on the causes of these. In earlier stages, gastric lesions are associated with few systemic symptoms, so it is the most praiseworthy thing to do in case of any slightest discomfort to go to the family physician. Severe gastric discomfort as well as gastric neoplasia is also described in the literature reviewed, it is more frequent in men after the age of 50 and increases with age.

PERCENTAGE OF PATIENTS CLASSIFIED BY EDUCATIONAL LEVEL WHO UNDERWENT UPPER GASTROINTESTINAL ENDOSCOPY. SOLCA CHIMBORAZO. SEPTEMBER 2012 - JANUARY 2013.

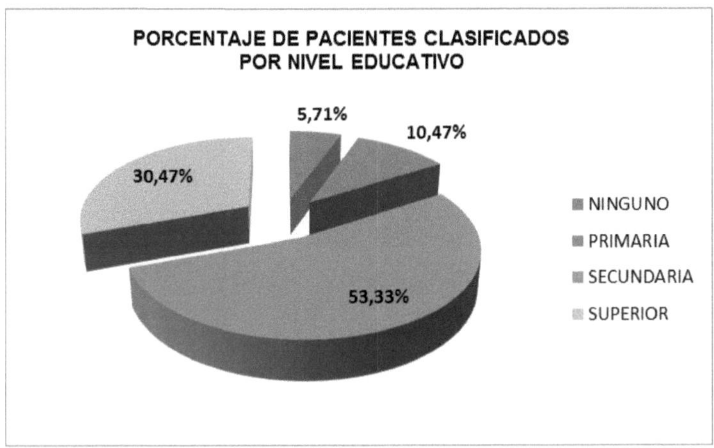

Graph No. 3. Patients classified by educational level who underwent upper gastrointestinal endoscopy. Solca-Chimborazo between September 2012 and January 2013.

This table indicates that most of the patients who underwent upper endoscopy correspond to the group whose education is secondary with 53.33%, followed by higher education with 30.47%, also establishing an existing correlation with gastric problems. According to the levels of stress and the hectic life they may lead, it is also necessary to consider the socioeconomic factors, since it is accepted that premalignant and malignant lesions are more frequent in people of lower socioeconomic level.

PERCENTAGE OF PATIENTS CLASSIFIED BY PLACE OF ORIGIN WHO UNDERWENT UPPER GASTROINTESTINAL ENDOSCOPY. SOLCA CHIMBORAZO. SEPTEMBER 2012 - JANUARY 2013.

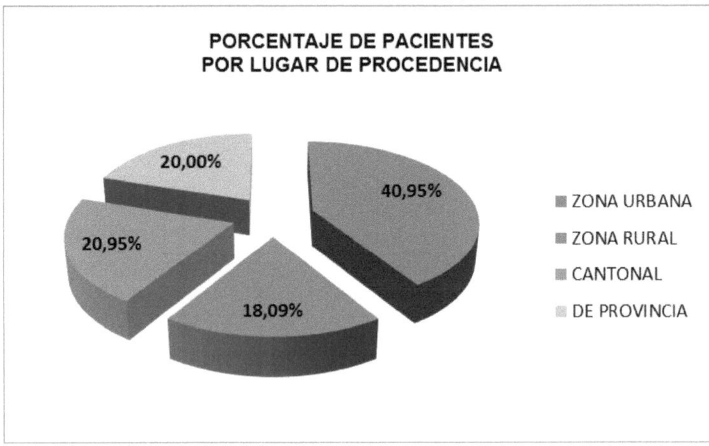

Graph No. 4.   **Patients classified by place of origin who underwent upper gastrointestinal endoscopy. Solca-Chimborazo between September 2012 and January 2013.**

It is evident that these territorial differences not only attribute to reasons related to the quality of the diagnosis and the process applied by each professional, but are influenced by a series of risk factors that differ in the diverse populations of our country. Patients in the urban area with 40.95 %, together with the cantonal area with 20.95 %, have almost the same conditions of gastrointestinal discomfort.

PERCENTAGE OF PATIENTS CLASSIFIED BY JOB PERFORMED WHO UNDERWENT UPPER GASTROINTESTINAL ENDOSCOPY. SOLCA CHIMBORAZO. SEPTEMBER 2012 - JANUARY 2013.

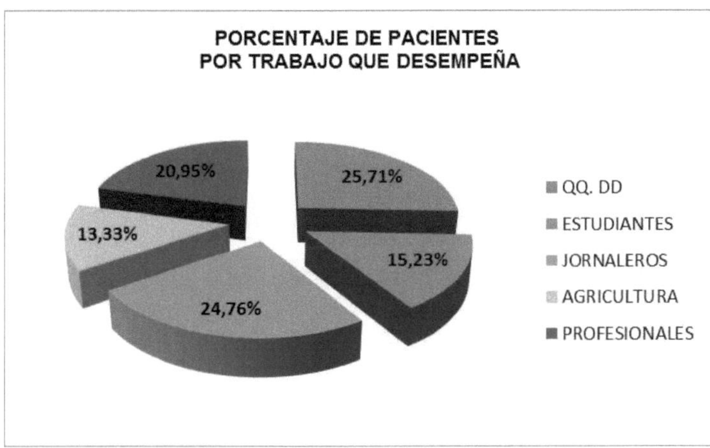

Graph No. 5.   **Patients classified by job performed who underwent upper gastrointestinal endoscopy. Solca-Chimborazo between September 2012 and January 2013.**

It has been shown that work stress influences biliary reflux, which is observed in this graph, since 59.04 % of the total are workers, with respect to 25.71 % of patients who are engaged in domestic chores and 25.71 % of students. Therefore, the work component influences the appearance of gastric diseases.

PERCENTAGE OF PATIENTS WHO HAD POSITIVE NITRITES DETERMINED IN GASTRIC JUICE. UPPER GASTROINTESTINAL ENDOSCOPY EXAMINATION. SOLCA CHIMBORAZO. SEPTEMBER 2012 - JANUARY 2013.

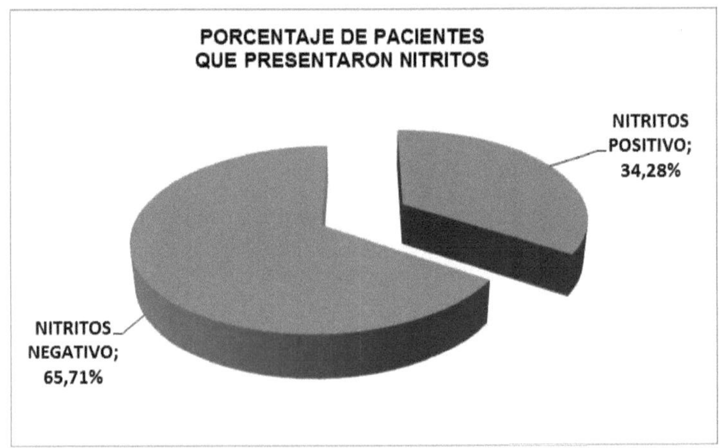

Graph No. 6.   Patients who showed the presence of nitrites in gastric juice who performed upper gastrointestinal endoscopy. Solca-Chimborazo between September 2012 and January 2013.

This graph shows the presence of nitrites in 36 patients (34.28%).

% of the total with respect to the 65.71 % of patients whose gastric acidity was normal. This result is interesting and therefore the cause of the existence of these substances should be investigated, however, the attempts to identify the elements of the diet related to the disease and the intervention studies carried out with these elements have been scarcely fruitful. This is not surprising for two reasons: 1) the difficulty and complexity of measuring the exact composition of these elements, since the same food can contain products with opposing effects; 2) diet is one more environmental factor that must be evaluated in the patient's environment and in the context of genetic susceptibility, which would require a more detailed estimate of its impact on lesions of the digestive tract.

PERCENTAGE OF PATIENTS WHO HAD HYPOCHLORHYDRIA OR ACHLORHYDRIA DETERMINED IN GASTRIC JUICE. UPPER GASTROINTESTINAL ENDOSCOPY EXAMINATION. SOLCA CHIMBORAZO. SEPTEMBER 2012 - JANUARY 2013.

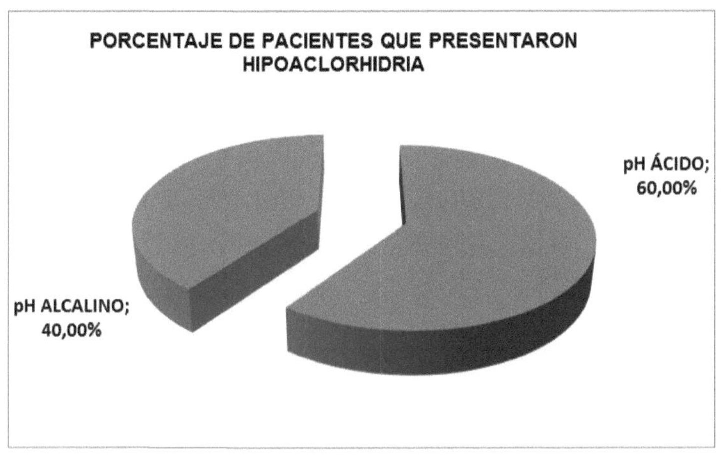

Patients who showed hypochlorhydria in gastric juice who underwent upper gastrointestinal endoscopy. Solca-Chimborazo between September 2012 and January 2013.

This graph shows that of the patients seen in the endoscopy department during the study period, 60% presented a low pH, which is normal; however, 40% had an alkaline pH, which indicates that these patients suffer from achlorhydria. It is known that the presence of reflux, *Helicobacter pylori* or both in the stomach causes an inflammatory process in the gastric mucosa that evolves in several stages, which depends on the characteristics of the reflux, the density and pathogenicity of the bacteria. This is why some authors point out that the histological damage observed in biopsy specimens in patients with gastric reflux with or without the presence of *Helicobacter pylori* may present different characteristics.

PERCENTAGE OF PATIENTS CLASSIFIED BY SEX WHO HAD PATHOLOGIES RELATED TO pH - NITRITES. UPPER GASTROINTESTINAL ENDOSCOPY EXAMINATION. SOLCA CHIMBORAZO. SEPTEMBER 2012 - JANUARY 2013.

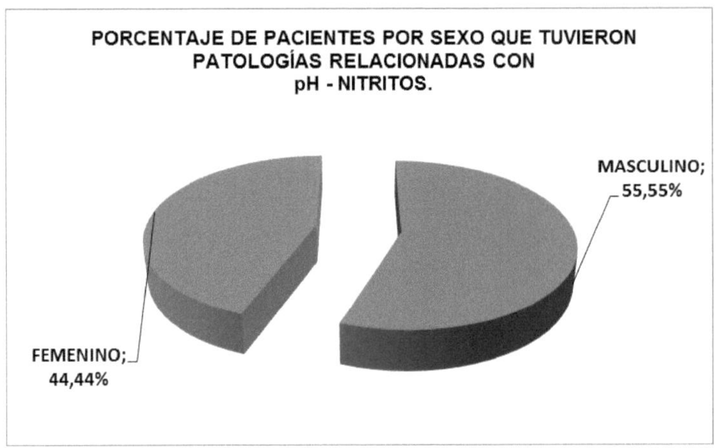

Graph No. 8.   **Patients who showed pH-nitrite relationship who underwent upper endoscopy. Solca-Chimborazo between September 2012 and January 2013.**

This graph shows that the difference in the pH-nitrite ratio between the two sexes is not so marked; the male with 55.55 % and the female with 44.44 %.

% of this relationship as stated above involving certain external factors of bad health habits can make heartburn disappear.

NUMBER OF PATIENTS CLASSIFIED BY SEX WHO HAD COMPLETE AND INCOMPLETE INTESTINAL METAPLASIA DIAGNOSED IN BIOPSIES TAKEN IN UPPER ENDOSCOPY EXAMINATION. SOLCA CHIMBORAZO. SEPTEMBER 2012 - JANUARY 2013.

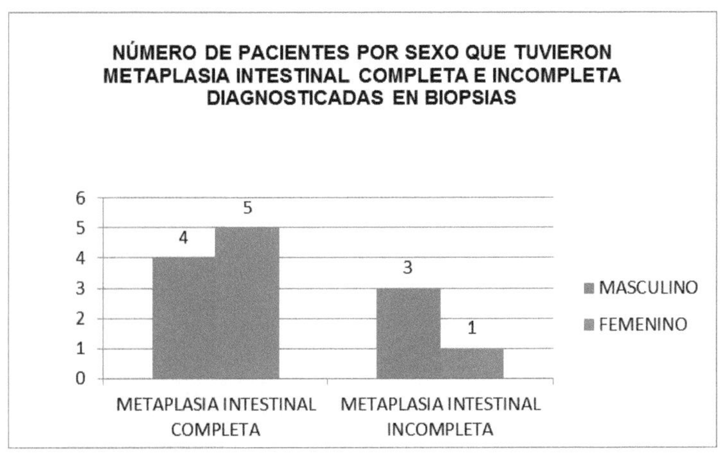

Graph No. 9 Patients classified by sex who had complete and incomplete intestinal metaplasia diagnosed in biopsy who underwent upper endoscopy. Solca-Chimborazo between September 2012 and January 2013.

In the period studied, of the 36 patients evaluated, 13 patients were histologically diagnosed with intestinal metaplasia according to the complete and incomplete classification; in these pathologies it is not indicated whether they are focal or diffuse. Intestinal metaplasia was more frequently observed in the gastric antrum, coinciding with a higher proportion of nitrite positivity. In these patients there does not seem to be an association between the presence of *Helicobacter pylori* and the intense degrees of intestinal metaplasia, in different studies performed it tells us that the presence of intestinal metaplasia with or without H. pylori have 6.5 times more risk of developing gastric cancer. There are data suggesting that gastric lesions and metaplasia arise from different cell lineages, so that metaplasia may be a precursor lesion, but rather a marker of increased risk.

NUMBER OF PATIENTS CLASSIFIED BY SEX WHO HAD FIBROSIS AND DECREASE OF GLANDULAR GROUPS DIAGNOSED IN BIOPSIES TAKEN IN UPPER ENDOSCOPY EXAMINATION. SOLCA CHIMBORAZO. SEPTEMBER 2012 - JANUARY 2013.

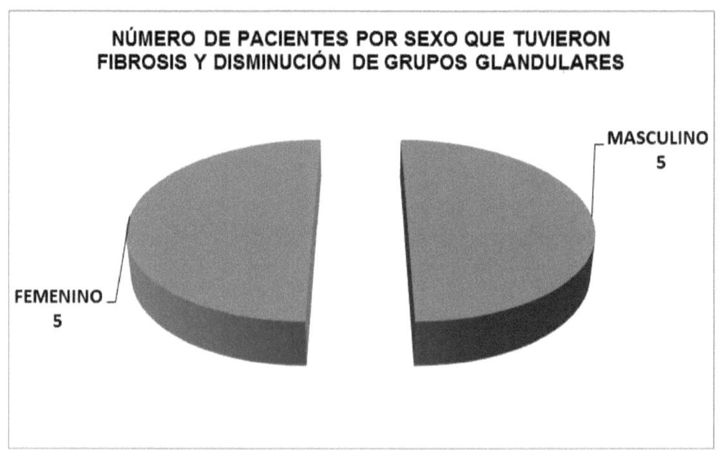

Graph No. 10 Patients who had fibrosis and decrease of glandular groups diagnosed in biopsy who underwent upper endoscopy. Solca- Chimborazo between September 2012 and January 2013.

Acute and chronic inflammatory infiltrate, as well as severe inflammatory infiltrate is associated with the decrease of the epithelium glands. It can be said to be the cause of the presence of these pathologies. This graph shows that out of the total of 36 samples analyzed, 10 patients, i.e. a quarter present fibrosis and decreased glandular clusters. Certain risk factors are common to other forms of lesions, so, despite the lack of direct evidence on the effect it could have on the incidence of the presence of other substances in the gastric mucosa, exposure to them should be limited, encouraging a healthy diet, increasing the consumption of fruits and vegetables, reducing fats and salt or foods preserved in it, practicing physical activity and not smoking.

NUMBER OF PATIENTS CLASSIFIED BY SEX WHO HAD INTESTINAL METAPLASIA WITH FIBROSIS AND DECREASE OF GLANDULAR GROUPS DIAGNOSED IN BIOPSIES TAKEN IN UPPER GASTROINTESTINAL ENDOSCOPY EXAMINATION. SOLCA CHIMBORAZO. SEPTEMBER 2012 - JANUARY 2013.

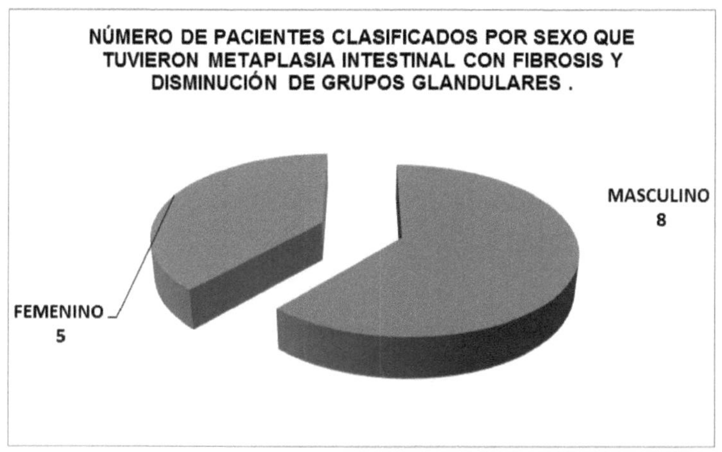

Patients who presented intestinal metaplasia with fibrosis and decrease of glandular groups diagnosed in biopsy who underwent upper gastrointestinal endoscopy. Solca-Chimborazo between September 2012 and January 2013.

The concordance studies carried out between the patient's clinic, the endoscopic diagnosis and the histopathological confirmation have shown that a gastric lesion (atrophy - metaplasia) could be diagnosed, even from the symptoms, in this graph shows that the lesion covers men in greater proportion, 8 patients with respect to 5 women, so it can be associated to more than presenting a decrease in acidity, it can also be correlated with other external factors such as the consumption of tobacco, alcohol etc, which add up to a severe irritation to the digestive tract.

5. CONCLUSIONS AND RECOMMENDATIONS

5.1 CONCLUSIONS

- Of a total of 105 patients analyzed who underwent upper gastrointestinal endoscopy, 100% presented clinical signs of gastrointestinal disease, including abdominal pain, nausea, vomiting, discomfort when digesting food, intermittent stomach pain and reflux. It is important to note that there were patients who presented more than one clinical sign.

- According to the results obtained, the presence of nitrites was observed in 36 patients (34.28% of the total), and 42 patients presented achlorhydria (40%).

- It was found that 36 patients out of the total had pathologies related to pH-nitrites, 55.55% of them were male and 44.44% female.

- In the period studied, of the 36 patients evaluated, 13 were histologically diagnosed with intestinal metaplasia according to the complete and incomplete classification, which was most frequently observed in the gastric antrum, coinciding with a higher proportion of nitrite positivity. In addition, 10 patients presented fibrosis and decreased glandular clusters.

- It was possible to determine the patients according to sex, age, profession and place of residence that there was a relationship between the macroscopic findings and the histopathological result, and the usefulness of the endoscopic procedure. In addition, gastric diseases can occur at any age and sex.

- The elevation of pH in gastric juice due to the disappearance of parietal cells was one of the main risk factors responsible for the development of intestinal metaplasia and even more so with the presence of nitrites and achlorhydria.

5.2 RECOMMENDATIONS

- All patients undergoing digestive endoscopy (upper, lower or both) must fall into the category of gastrointestinal patients, to whom a diagnostic plan must be postulated that should include: complete anamnesis, clinical examination, complete blood test, biochemical profile, urinalysis, coproparasitic, in order to rule out diseases that as secondary signs present gastrointestinal conditions such as parasitic, infectious, hepatic, renal and other diseases.

- It should be recommended and taken into account that digestive endoscopy is not only diagnostic, but can also be therapeutic as in cases of foreign body where, depending on the size and location, it can be removed without the need to submit the patient to a surgical procedure.

- Patient information and education is basically a relational process and is, therefore, a verbal process, in which there is a continuous interaction and exchange of information between the healthcare professional and the patient. It could also be said that the information criterion to be applied in the clinical relationship is always subjective. Each patient should be provided with all the information that, given his or her personal circumstances, he or she needs to make a decision.

- Endoscopy is a technique that makes lesions visible and is able to provide morphological and biochemical information as long as biopsies are taken from the possible lesions that may exist.

- There is sufficient epidemiological evidence to affirm that diet plays an important role in the development and prevention of gastric metaplasia and atrophy, gastric or colorectal cancer. In addition, it has been possible to identify dietary patterns or behaviors that would explain a good part of the interregional variability that exists in the incidence and mortality of some gastric pathologies and that would facilitate public health recommendations, the efficacy of which should be evaluated in the future.

6. BIBLIOGRAPHY .

1. CORREA P, COELLO C, DUQUE (1999). INTESTINAL CARCINOMA AND METAPLASIA OF THE STOMACH IN COLOMBIAN MIGRANTS. CANCER INSTITUTE; 44:297-306.

2. CORNEE J, LAIRON D, VELEMA J, GUYADER M, BERTHEZENE P (1992). AN ESTIMATE OF NITRATE, NITRITE AND N-NITROSODIMETHLYAMINE CONCENTRATIONS IN FRENCH FOOD PRODUCTS OR FOOD GROUPS.SCIENCES DES ALIMENTS, 12:155-97.

3. CHOTIPRASIDHI P. (2000). EFFECTIVENESS SINGLE DILATATION WITH MALONEY DILATOR VS ENDOSCOPIC RUPTURTE OFF SCHATZKI CAP. 45 PAG 28.

4. DALENBACK J. (1996). MECHANISMS BEHIND CHANGES IN GASTRIC ACID AND BICARBONATE OUTPUTS DURING THE HUMAN INTERDIGESTIVE MOTILITY CYCLE. CH 270 P 113.

5. DE WEERTH A, GOCHT A, SEEWALD S, BRAND B (2002), ET AL. DUODENAL NODULAR LYMPHOID HYPERPLASIA CAUSED BY GIARDIASIS INFECTION IN A PATIENT WHO IS IMMUNODEFICIENT. GASTROINTEST ENDOSC; 55(4):605-607.

6. DRUCKER R. (2005). FISIOLOGIA MÉDICA. MEXICO. EDITORIAL EL MANUAL MODERNO. 3ER ED.

7. DVORKIN, MA (2003). PHYSIOLOGICAL BASES OF MEDICAL PRACTICE 13TH EDITION. SPAIN. MEDICA PANAMERICANA.

8. FARRERAS-ROZMAN (2000). INTERNAL MEDICINE. 14TH EDITION

VOLUME I, HARCOURT PUBLISHING HOUSE; PG. 172.

9. GARCÍA - CONDE J.; MERINO J.; GONZALES J (1995). GENERAL PATHOLOGY, CLINICAL SEMIOLOGY AND PHYSIOPATHOLOGY. MC GRAW HILL INTERAMERICANA.

10. GARTNER L. P.; HIATT J. L. (1997) HISTOLOGY TEXT AND ATLAS. EDITORIAL MC GRAW-HILL INTERAMERICANA.

11. GUYTON A. (1999). TREATISE ON MEDICAL PHYSIOLOGY. NINTH EDITION. EDITORIAL MC GRAW-HILL INTERAMERICANA.

12. GREENBERGER N, BLUMBERG R, BURAKOFF R (2009). CURRENT DIAGNOSIS & TRATMENT. GASTROENTEROLOGY, HEPATOLOGY & ENDOSCOPY. NORTON J., MCGRAW HILL, 2009. PAGES. 64, 184, 185 Y 240.

13. HAN K, PEURA D. (2008). ASSOCIATION BETWEEN HELICOBACTER PYLORI INFECTION AND GASTROINTESTINAL MALIGNANCY.

14. HAWKEY CJ, LANGMAN MJ (2003). NON STEROIDAL ANTI-INFLAMMATORY DRUGS OVERALL RISK AND MANAGEMENT, P. 608.

15. HIB J. (1999) MEDICAL EMBRYOLOGY. SEVENTH EDITION. PAG. 268. EDITORIAL MC GRAW-HILL. INTERAMERICANA.

16. INEN 2012. ECUADORIAN INSTITUTE OF STATISTICS AND CENSUS.

17. JM SANZ ANQUELA, BLASCO M. (2005). GASTRIC PATHOLOGY: PRECURSOR LESIONS OF GASTRIC CANCER. REVIEW.

CONFERENCE VII VIRTUAL HISPANIC-AMERICAN CONGRESS OF ANATOMIC PATHOLOGY. OCTOBER 2005.

18. L. ABREU (2006). GASTROENTEROLOGY: DIAGNOSTIC AND THERAPEUTIC ENDOSCOPY. 2ND ED. BUENOS AIRES. EDITORIAL MÉDICA PANAMERICANA. PG. XIII.

19. LATARJET M (2005) ANATOMÍA HUMANA TOMO II, 4TA ED. MEXICO EDITORIAL MÉDICA PANAMERICANA.

20. LESSON T. S.; LESSON C. R. (1980). SECOND EDITION . HISTOLOGY PAGE 206. EDITORIAL MÉDICA PANAMERICANA.

21. LIJIMA K, SHIMOSEGAWA T.(2006). GASTRIC CARDITIS: IS IT A HISTOLOGICAL RESPONSE TO HIGH CONCENTRATIONS OF LUMINAL NITRIC OXIDE? WORLD J GASTROENTEROL. SEP 28-12 (36):5767-71.

22. NAYLOR G, AXON A., (2003). ROLE OF BACTERIAL OVERGROWTH IN THE STOMACH AS AN ADDITIONAL RISK FACTOR FOR GASTRITIS, CAN J GASTROENTEROL. JUN-17 SUPPL B13, B-17B.

23. ODZE R., GOLDBLUM J., (2009). SURGICAL PATHOLOGY OF THE GI TRACT, LIVER, BILIARY TRACT, AND PANCREAS, SANDERS, ELSEVIER. PAG. 293 Y 294.

24. PÉREZ E. ABDO J. BERNAL F. (2012). GASTROENTEROLOGY. MEXICO ED. MC GRAWHILL. PAG. 145.

25. POPE. C. (1993). NORMAL ANATOMY AND DEVELOPMENTAL ANOMALIES. 5TH ED PHIDELPHIA. PAGE 311-318

26. RALT D AND TANNENBAUM SR (1981). THE ROLE OF BACTERIA IN NITROSAMINE FORMATION. IN: NNITROSO COMPOUNDS, R.A. SCANLAN AND S.R. TANNENBAUM (EDS.), ADVANCES IN CHEMISTRY SERIES NO. 174, AMERICAN CHEMICAL SOCIETY: WASHINGTON, DC,;PP. 157-164

27. RAMIREZ-RAMOS A, GILMAN R (2004). HELICOBACTER PYLORI IN PERU. LIMA-PERU. 2004. EDITORIAL SANTA ANA S.A. (276 PP.).

28. RAMOS N F. (2000). GASTROPATHIES PRODUCED BY NON-STEROIDAL INFLAMMATORY AGENTS. MEDICINA UNIVERSITARIA 3(9); EDITORIAL EDAMSA. IMPRESIONES. MEXICO PAG 2.

29. ROJO, J. (2003). NEW THERAPIES IN THE MANAGEMENT OF CHRONIC INFLAMMATORY BOWEL DISEASE. EDITORIAL MC GRAW HILL INTERAMERICANA.

30. ROUVIEREH . (2001). ANATOMÍAHUMANA . DESCRIPTIVE, TOPOGRAPHIC AND FUNCTIONAL. VOLUME II. 10TH EDITION. P 316, 348-
349. ED. MASSON. PARIS.

31. SUGIMURA, T. (2000). NUTRITION AND DIETARY CARCINOGENS. CARCINOGENESIS; 21: 387-195.

32. SUZUKI H, K IIJIMA, A MORIYA, K MCELROY, G SCOBIE, V FYFE AND K E L MCCOLL ARE (2003). MAXIMAL AT THE GASTRIC CARDIA CONDITIONS FOR ACID CATALYSED LUMINAL NITROSATION.

PAG;:1095.

33. SCHWARTZ (2005). PRINCIPLES OF GENERAL SURGERY. VOL II 8TH ED, MEXICO MCGRAW- HILL INTERAMERICANA.

34. SLEISENGER - FORDTRAN (2002). GASTROINTESTINAL AND LIVER DISEASES. 7MA ED. EDITORIAL MÉDICA PANAMERICANA. PRINTED IN ARGENTINA. PAG. 542-543; 717 CHAPS. 36, 757, 760.

35. SPITZ L. (1996). ESOPHAGEAL ATRESIA. PAST, PRESENT AND FUTURE. SURG 31:19.

36. WALTERS CL. (1992). REACTIONS OF NITRATE AND NITRITE IN FOODS WITH SPECIAL REFERENCE TO THE DETERMINATION OF N-NITROSO COMPOUNDS. FOOD ADDIT CONTAM; CHAP. 9. PAG. 441.

37. WINK DA, FELLISCH M, VODOVOTZ Y, (1999) ET AL. IN: GILBERT DL, COTON CA, EDS.REACTIVE OXYGEN SPECIES IN BIOLOGICAL SYSTEMS. NEW YORK: KLUWER. ACADEMIC/PLENUM PUBLISHERS. : 245-91.

38. WOLFE, M.M. (2002). THERAPEUTICS OF DIGESTIVE DISORDERS. EDITORIAL MC GRAW HILL INTERAMERICANA.

39. YAZBECK C. (1995). GASTROINTESTINAL EMERGENCIES OF THE NEONATE. 4TH ED. ST. LOUIS, MOSBIC - YEARBOOK. PAG. 53.

40. ZHANG C (2005). HELICOBACTER INFECTION, GLANDULAR ATROPHY AND INTESTINAL METAPLASIA IN SUPERFICIAL GASTRITIS GASTRIC EROSIVE GASTRITIS, GASTRIC ULCER AND AERLY GASTRIC CANCER. WORLD GASTROENTEROL. PAG 791.

7. ANNEXES

ANNEX No. 1 RESULTS SHEET OF THE 36 PATIENTS WITH HYSTOPATHOLOGICAL DIAGNOSIS ,
atrophy, gastric metastasis, achlorhydria and nitrites at SOLCA CHIMBORAZO HOSPITAL sep. 2012 - jan. 2013.

Pcte. No.	AGE	SEX	NITRITES	pH Aclorhidria	1	2	3
1	22	F	POSITIVE	POSITIVE	X		
2	64	M	POSITIVE	POSITIVE		X	
3	35	M	POSITIVE	POSITIVE	X		
4	33	M	POSITIVE	POSITIVE			X
5	44	F	POSITIVE	POSITIVE	X		
6	24	F	POSITIVE	POSITIVE			X
7	35	M	POSITIVE	POSITIVE		X	
8	46	M	POSITIVE	POSITIVE			X
9	31	M	POSITIVE	POSITIVE			X
10	72	F	POSITIVE	POSITIVE		X	
11	28	F	POSITIVE	POSITIVE			X
12	44	M	POSITIVE	POSITIVE	X		
13	58	F	POSITIVE	POSITIVE		X	
14	36	M	POSITIVE	POSITIVE			X
15	26	F	POSITIVE	POSITIVE			X
16	44	M	POSITIVE	POSITIVE			X
17	31	F	POSITIVE	POSITIVE	X		
18	30	M	POSITIVE	POSITIVE			X
19	66	M	POSITIVE	POSITIVE	X		
20	40	F	POSITIVE	POSITIVE			X
21	21	M	POSITIVE	POSITIVE	X		
22	56	F	POSITIVE	POSITIVE		X	
23	37	F	POSITIVE	POSITIVE			X
24	71	F	POSITIVE	POSITIVE	X		
25	39	M	POSITIVE	POSITIVE			X
26	42	M	POSITIVE	POSITIVE	X		
27	70	F	POSITIVE	POSITIVE		X	
28	41	M	POSITIVE	POSITIVE	X		
29	35	F	POSITIVE	POSITIVE			X
30	68	M	POSITIVE	POSITIVE	X		
31	49	M	POSITIVE	POSITIVE		X	
32	67	F	POSITIVE	POSITIVE		X	
33	62	F	POSITIVE	POSITIVE	X		
34	55	M	POSITIVE	POSITIVE		X	
35	42	M	POSITIVE	POSITIVE		X	
36	77	M	POSITIVE	POSITIVE	X		

LEGEND

1. METAPLASIA

2. FIBROSIS AND DECREASE OF GLANDULAR CLUSTERS

3. METAPLASIA-DIMINUTION OF GLANDULAR CLUSTERS

Pcte. No.	AGE	SEX	NITRITES	pH Aclorhidria
37	26	M	NEGATIVE	NEGATIVE
38	52	F	NEGATIVE	NEGATIVE
39	36	M	NEGATIVE	NEGATIVE
40	26	F	NEGATIVE	NEGATIVE
41	44	F	NEGATIVE	NEGATIVE
42	31	M	NEGATIVE	NEGATIVE
43	30	F	NEGATIVE	NEGATIVE
44	66	M	NEGATIVE	NEGATIVE
45	40	F	NEGATIVE	NEGATIVE
46	21	F	NEGATIVE	NEGATIVE
47	28	M	NEGATIVE	NEGATIVE
48	39	F	NEGATIVE	NEGATIVE
49	42	M	NEGATIVE	NEGATIVE
50	70	M	NEGATIVE	NEGATIVE
51	41	F	NEGATIVE	NEGATIVE
52	35	F	NEGATIVE	NEGATIVE
53	68	M	NEGATIVE	NEGATIVE
54	49	F	NEGATIVE	NEGATIVE
55	67	M	NEGATIVE	NEGATIVE
56	62	F	NEGATIVE	NEGATIVE
57	55	F	NEGATIVE	NEGATIVE
58	42	M	NEGATIVE	NEGATIVE
59	77	F	NEGATIVE	NEGATIVE
60	71	M	NEGATIVE	NEGATIVE
61	28	F	NEGATIVE	NEGATIVE
62	35	F	NEGATIVE	NEGATIVE
63	68	F	NEGATIVE	NEGATIVE
64	49	F	NEGATIVE	NEGATIVE
65	67	M	NEGATIVE	NEGATIVE
66	62	F	NEGATIVE	NEGATIVE
67	42	F	NEGATIVE	NEGATIVE
68	25	M	NEGATIVE	NEGATIVE
69	55	M	NEGATIVE	NEGATIVE
70	36	F	NEGATIVE	NEGATIVE
71	28	F	NEGATIVE	NEGATIVE
72	69	M	NEGATIVE	NEGATIVE

Pcte. No.	AGE	SEX	NITRITES	pH Chlorhydria
73	22	F	NEGATIVE	NEGATIVE
74	64	M	NEGATIVE	NEGATIVE
75	35	F	NEGATIVE	NEGATIVE
76	33	F	NEGATIVE	NEGATIVE
77	44	F	NEGATIVE	NEGATIVE
78	58	F	NEGATIVE	NEGATIVE
79	36	M	NEGATIVE	NEGATIVE
80	26	F	NEGATIVE	NEGATIVE
81	44	M	NEGATIVE	NEGATIVE
82	31	F	NEGATIVE	NEGATIVE
83	30	F	NEGATIVE	NEGATIVE
84	66	M	NEGATIVE	NEGATIVE
85	40	F	NEGATIVE	NEGATIVE
86	21	F	NEGATIVE	NEGATIVE
87	56	M	NEGATIVE	NEGATIVE
88	44	F	NEGATIVE	NEGATIVE
89	31	F	NEGATIVE	NEGATIVE
90	30	M	NEGATIVE	NEGATIVE
91	66	F	NEGATIVE	NEGATIVE
92	40	M	NEGATIVE	NEGATIVE
93	21	F	NEGATIVE	NEGATIVE
94	56	F	NEGATIVE	NEGATIVE
95	37	F	NEGATIVE	NEGATIVE
96	71	F	NEGATIVE	NEGATIVE
97	39	M	NEGATIVE	NEGATIVE
98	42	F	NEGATIVE	NEGATIVE
99	70	M	NEGATIVE	NEGATIVE
100	41	F	NEGATIVE	NEGATIVE
101	35	M	NEGATIVE	NEGATIVE
102	27	F	NEGATIVE	NEGATIVE
103	36	F	NEGATIVE	NEGATIVE
104	28	F	NEGATIVE	NEGATIVE
105	35	F	NEGATIVE	NEGATIVE

APPENDIX No. 3RECORD OFDEFICENTIFICATIONDEFICFICATIONS OF PATIENTS SEEN IN THE ENDOSCOPY DEPARTMENT. HOSPITAL SOLCA CHIMBORAZO SEPTEMBER 2012 - JANUARY 2013.

Clinical record No.	Date:
Patient's name:	
Place of residence	
Age:	
Sex:	
Clinical signs:	
Examination requested:	
Endoscopic findings:	
Presumptive diagnosis:	
Histopathological diagnosis:	

ANNEX No. 4 REPORT MODEL OF THE REPORT OF THE DEPARTMENT OF ENDOSCOPY. SOLCA CHIMBORAZO HOSPITAL SEPTEMBER 2012 - JANUARY 2013.

<table>
<tr>
<td rowspan="2">HOSPITAL ONCOLOGICO
DR. FAUSTO ANDRADE YANEZ
UNIDAD ONCOLOGICA SOLCA CHIMBORAZO</td>
<td rowspan="2">INFORME DE ENDOSCOPIA
DIGESTIVA ALTA</td>
<td>SERVICIO DE</td>
</tr>
<tr>
<td>ENDOSCOPIA</td>
</tr>
<tr>
<td>NOMBRE:</td>
<td>EDAD:
56 a</td>
<td>H.C. Nº:</td>
</tr>
<tr>
<td colspan="2">MÉDICO SOLICITANTE: Dr.</td>
<td>ORIGEN: (Comunidad Lirio San José) GUAMOTE</td>
</tr>
<tr>
<td colspan="3">SÍNTOMAS: Dolor, prominencia en la boca del estómago.</td>
</tr>
<tr>
<td colspan="3">IDG:</td>
</tr>
<tr>
<td>EQUIPO:
OLIMPUS EXERA GIF V.</td>
<td colspan="2">PREMEDICACIÓN:
Midazolam 2 mg.</td>
</tr>
<tr>
<td colspan="3">ESOFAGO:
de fácil acceso, de características normales.</td>
</tr>
<tr>
<td colspan="3">ESTOMAGO:
luz y calibre disminuído, lago mucoso en moderada cantidad verde de orígen biliar.
Lesión ulcero-infiltrativa que se extiende desde incisura angulares hasta región prepilórica que impide la visualización del píloro.
La lesión tiene una extensión de unos 5 x 3 cm que ocupa curvatura menor del antro y parte de paredes anterior y posterior.
Se biopsia con signo de tienda de campaña negativo.</td>
</tr>
<tr>
<td colspan="3">DUODENO:
no se explora por dificultad de píloro excéntrico forzado</td>
</tr>
<tr>
<td colspan="3">UREASA:</td>
</tr>
<tr>
<td colspan="3">PATOLOGÍA:</td>
</tr>
<tr>
<td colspan="3">IDGE:
LESIÓN ULCERO-INFILTRATIVA DE CURVATURA MENOR DE ANTRO TIPO BORMANN III, COMPATIBLE CON ADENO CA.</td>
</tr>
<tr>
<td colspan="2">Dr. Fabián Romero R</td>
<td>Fecha:
/10/2013</td>
</tr>
</table>

'SOLCA' UNA MANO AMIGA AL SERVICIO DE LOS CHIMBORACENSES

ANNEX No. 5 REPORT OF HISTOPATHOLOGICAL REPORT DIAGNOSED IN THE DEPARTMENT OF PATHOLOGY. HOSPITAL SOLCA CHIMBORAZO SEPTEMBER 2012 - JANUARY 2013.

SOLCA UNIDAD ONCOLÓGICA SOLCA CHIMBORAZO	INFORME ANÁTOMO PATOLÓGICO	LABORATORIO DE PATOLOGÍA
		Nº. INFORME 31130 EXTERNOS

```
: PACIENTE :                              SEXO : M EDAD : 51 Años   H.Cl.:                    00018351 :
: ESTABLECIMIEN/CIUDAD : CEBADAS            DIRECCION :                          TELF :               :
: Nº HIS.CLI. REFEREN. :                    FECHA REF.:                                               :
: SERVICIO REMITENTE    : Gastroenterología FECHA PED.:              INF.CITOLOGICO :                 :
: MEDICO SOLICITANTE    : DR. CARDENAS                                                                :
: TIPO DE LA MUESTRA    : Biopsia Incisional                                                          :
: ORIGEN DE LA MUESTRA  : MUCOSA GASTRICA                                                             :
: INFORMACION CLINICA   :                                                                             :
:                                                                                                     :
: PRESUNCION CLINICA    :                                                                             :
: INFORMES PREVIOS      :                                                                             :

: HISTOQUIMICA : 0    INMUNOHISTOQ. : 0    FOTOGRAFIA : N    Nº PLACAS : 1    Nº MUESTRAS : 1    ARCHIVAR :12m:
```

INFORME:

MACROSCOPIA:

Se recibe por separado:

1.ROTULADO "ANTRO".- Se recibe dos fragmentos de tejido irregular de consistencia blanda de color gris blanquecino que en conjunto miden 0,4cm. Se procesa toda la muestra.

2.ROTULADO "CUERPO".- Se recibe dos fragmentos de tejido irregular de consistencia blanda de color gris blanquecino que en conjunto miden 0,4cm. Se procesa toda la muestra.

3.ROTULADO "INCISURA".- Se recibe un fragmento de tejido irregular de consistencia blanda de color gris blanquecino que mide 0,3cm. Se procesa toda la muestra.

MICROSCOPIA:

DIAGNOSTICO:

DIAGNOSTICO DESCRIPTIVO
BIOPSIA DE MUCOSA GASTRICA

- Gastritis Crónica Erosiva Folicular Activa Moderada Atrófica, Focal Metaplasia Intestinal Completa
- Helicobacter Pylori (++)
- Grupo II

GA

TOPOGRAFIAS:
211.1 ESTOMAGO   535.0 GASTRITIS CRONICA ACTIVA.
000.4 HELICOBACTER PYLORI
MORFOLOGIAS:

SOLCA - Chimborazo, Lunes 30 de Septiembre de 2013   09:08:45

3 0 SEP 2013

PATOLOGIA

SRA. GABRIELA ALOMIA
PATOLOGO
Jefe de servicio

APPENDIX No. 6 GASTRIC PHYSIOLOGY

ANOMALIA	INCIDENCIA	EDAD DE PRESENTACION	SIGNOS Y SINTOMAS	TRATAMIENTO
Atresia gástrica, antral o pilórica	3:1000.000, cuando se combinan dos membranas	Lactancia	Vómitos no biliosos	Gastroduodenostomia, gastroyeyunostomia
Membrana pilórica o antral	Igual que la anterior	Cualquier edad	Fracaso de la maduración, vómitos	Incisión o exeresis piloroplastia
Microgastria	Rara	Lactancia	Vómitos, desnutrición	Alimentos por goteo continuo o bolsillo yeyunal como reservorio
Estenosis pilórica	En los Estados Unidos, 3:1.000 (regional, 1:1.000-8:1.000) masculino/femenino 4:1	Lactancia	Vómitos no biliosos	Piloromiotomia
Duplicación gástrica	Rara masculino/femenino, 1:2	Cualquier edad	Masa abdominal, vómitos, hematemesis, peritonitis si se rompe	Exeresis o gastrectomía parcial
Divertículo gástrico	Rara	Cualquier edad	Por lo general, asintomático	No suele ser necesario
Teratoma gástrico	Rara	Cualquier edad	Masa abdominal alta	Resección
Vólvulo gástrico	Rara	Cualquier edad	Vomitos, rechazo del alimento	Reducción del vólvulo, gastropexia anterior
Atresia o estenosis duodenal	1:20.000	Neonato	Vómitos biliosos, distensión abdominal alta	Duodenoyeyunostomia o gastroyeyunostomia
Páncreas anular	1:10.000	Cualquier edad	Vómitos biliosos, fracaso de la maduración	Duodenoyeyunostomia
Duplicación duodenal	Rara	Cualquier edad	Hemorragia digestiva, dolor	Exeresis
Malrotacion y vólvulo del intestino medio	Rara	Cualquier edad	Vómitos biliosos, distensión abdominal alta	Reducción, sección de bandas, posiblemente resección.

Printed by Books on Demand GmbH, Norderstedt / Germany